NATURAL AND ANTHROPOGENIC IMPACTS ON CAVE CLIMATES

NATURAL AND ANTHROPOGENIC IMPACTS ON CAVE CLIMATES

Postojna and Predjama Show Caves of Slovenia

STANKA ŠEBELA
ZRC SAZU
Karst Research Institute
Postojna
Slovenia

ELSEVIER

Elsevier
Radarweg 29, PO Box 211, 1000 AE Amsterdam, Netherlands
The Boulevard, Langford Lane, Kidlington, Oxford OX5 1GB, United Kingdom
50 Hampshire Street, 5th Floor, Cambridge, MA 02139, United States

Library of Congress Cataloging-in-Publication Data
A catalog record for this book is available from the Library of Congress

British Library Cataloguing-in-Publication Data
A catalogue record for this book is available from the British Library

ISBN: 978-0-12-822954-5

For information on all Elsevier publications visit our website at
https://www.elsevier.com/books-and-journals

Publisher: Candice Janco
Acquisitions Editor: Louisa Munro
Editorial Project Manager: Allison Hill
Production Project Manager: Sruthi Satheesh
Cover Designer: Victoria Pearson

Typeset by TNQ Technologies

Contents

Acknowledgments

Research data used for this manuscript were collected and analyzed within the research program *Karst Research* (P6-0119) and infrastructure program *Natural and Cultural Heritage* (I0-0031) financed by the Slovenian Research Agency. Writing of this manuscript was part of the research project *Karst Research for Sustainable Use of Škocjan Caves as World Heritage* (L7-8268) co-financed by the Slovenian Research Agency and Park Škocjanske Jame, and research project *Methodology for monitoring the sustainable use of karst show caves with automatic measurements—role model—Postojna cave* (L6-9397) co-financed by the Slovenian Research Agency, Postojnska jama d.d. and REODOM d.o.o.

Field work and the use of research infrastructure used for this monograph were performed within the project operations "Development of Research Infrastructure for the International Competitiveness of the Slovenian RRI Space—RI–SI–EPOS and RI–SI–LifeWatch" co-financed by the Republic of Slovenia, the Ministry of Education, Science and Sport, and the European Union from the European Regional Development Fund. Microclimatic monitoring in show caves was also part of EC Horizon2020 project EPOS SP.

I am grateful to the management of Postojnska Jama d.d. Company, Škocjan Caves Regional Park, and the management of Županova Jama and Kostanjeviška Jama for allowing me and my colleagues to perform microclimatic monitoring in their show caves. Thanks are also due to my co-workers at Karst Research Institute ZRC SAZU in Postojna where I have worked since 1988. This research was also part of the study programme at University of Nova Gorica (Karstology (third level) graduate school).

I thank also Allison Hill, Editorial Project Manager at Elsevier, and Peter W. Adamson, PhD Editor at Elsevier, who invited me to write this book.

I am grateful also to my family, especially my daughter Sara, for support during the writing of this book.

CHAPTER 1

Introduction

Slovenia is the home of Classical Karst, which is an area that extends across the border between southwest Slovenia (Figs. 1.1 and 1.2) and northeast Italy (Gams, 2004; Jurkovšek et al., 2016). The Classical Karst Region lies in the northwestern part of the External Dinarides, which stretch in a northwest–southeast direction (Jurkovšek et al., 2016). The international scientific term "karst" was adopted from the name of Classical Karst (Gams, 2003, 2004). Classical Karst is part of Dinaric Karst, which is the largest continuous karst surface in Europe and one of the largest karst surfaces in the world (https://whc.unesco.org/en/tentativelists/6072/).

In Slovenia, karst areas represent 43% of the country (see Fig. 5.1 in Chapter 5, Microclimatic monitoring in show caves). There are 13,660 registered caves in Slovenia (Cave Register, November 2020), what means that the cave density is 1.56 caves per 1 km^2 on average. A bare karst landscape was typical in the past but in the last 150 years vegetation has covered the karst areas and the expression "green karst" is becoming commonly used (Fig. 1.2). Nevertheless, there are still typical bare karst areas in southwest Slovenia with a strong northern wind called the "burja" or bora wind (Fig. 1.3).

Figure 1.1 Karst landscape in the area of Škocjan Caves with Vremščica mountain (1027 m) in the background. *(Photo: S. Šebela.)*

Natural and Anthropogenic Impacts on Cave Climates
ISBN 978-0-12-822954-5
https://doi.org/10.1016/B978-0-12-822954-5.00001-9

Figure 1.2 Karst landscape in the area of Škocjan Caves. *(Photo: S. Šebela.)*

Figure 1.3 Karst landscape illustrating the bora wind on the high plateau of Nanos Mountain (1313 m at the highest peak). *(Photo: S. Šebela.)*

This monograph primarily addresses microclimate research in two show caves in Slovenia (Postojna and Predjama Caves, Fig. 1.4). Those are also the caves with the longest air temperature data sets, as air temperature monitoring started at hourly intervals in 2009. Additionally, some other show caves (Škocjan Caves, Županova Jama, and Kostanjeviška Jama) in Slovenia are discussed more briefly (see Chapter 5, Microclimatic monitoring in show caves).

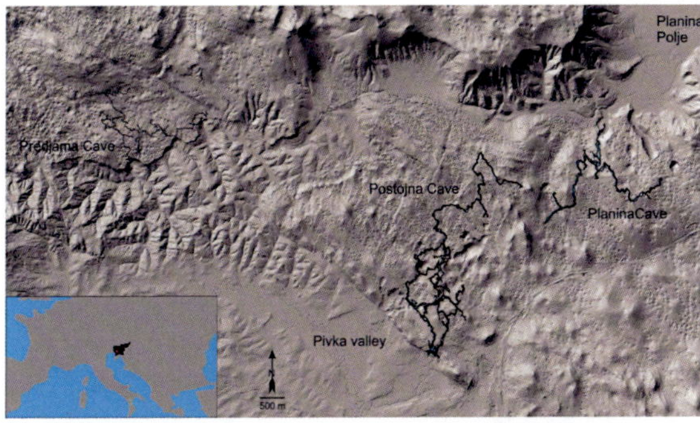

Figure 1.4 Geographic position of Postojna and Predjama Caves in southwest Slovenia (lidar relief from ARSO, 2020). On the small map of Europe Slovenia is marked with black, and the area of the Postojna and Predjama Caves is marked with a red dot.

Postojna Cave represents the southwest part of the Postojna-Planina Cave system (Fig. 1.4), because the underground water connection between both caves has been known since the explorations of Gruber (1781). There still remain about 800 m of unknown passages between Postojna and Planina Cave (6656 km long and 65 m deep) (Šebela, 2012, 2019). Postojna Cave is the second-longest cave in Slovenia (24,340 m long and 115 m deep). The cave was visited in prehistoric periods because cultural layers from Postojna Cave belong to the last half of the Middle Paleolithic (the end of the Mousterian) (Šebela, 2019). In the beginning of the Würm glacial period, the presence of Paleolithic man in the cave can be proved reliably (Brodar, 1966). Small portions of Postojna Cave were known in the 13th century (Gams, 2003; Čuk, 2008; Kranjc et al., 2007) (see Fig. 5.1.6.1 in Section 5.1.6, Visitors in Postojna Cave). However, the beginning of modern tourism started in the 19th century when the native Luka Čeč discovered new parts of the cave. The year 1818 is the official year of discovery of the most extensive section of the Postojna Cave and 1819 saw the beginning of modern tourism in Postojna Cave.

In a geological sense, Postojna Cave is developed in Cretaceous limestones (Šebela, 1998). Surface River Pivka sinks to 511 m at the contact of Eocene flysch rocks with limestone karst rocks (Fig. 1.5) and continues underground forming passages of Postojna Cave toward Planina Cave. The oldest cave sediments from Postojna Cave, that point to the age of cave passages, are dated to >2.58 and < 3.58 Ma (Zupan Hajna et al., 2020).

Postojna and Planina Caves are type localities for ~ 90 species of invertebrates. With more than 100 currently known aquatic and terrestrial

Figure 1.5 Ponor or sinking point of the River Pivka to Postojna Cave. *(Photo: S. Šebela.)*

troglobiotic species and a large number of troglophile and trogloxene aquatic and terrestrial species, the Postojna-Planina cave system is one of the richest cave systems in the world in terms of fauna. Postojna Cave is described as the birthplace of speleobiology, where the first specimens of cave animals were described (Culver and Pipan, 2014).

Predjama Cave is, according to the Cave Register, the third longest cave in Slovenia (16,755 m long and 168 m deep). Extensive speleological research in 2020, which has not been completed yet, indicates extension of the cave passages to ~17 km (Fidel, 2020), showing the complexity of incompletely discovered the cave system. Only an ~700 m long entrance part of Predjama Cave is accessible to tourists, with a vertical range of 50.74 m (Šebela and Turk, 2014a, b). The rest of the cave is suitable for trained speleologists only. In a geological sense, Predjama Cave is interesting because the passages run through Upper Cretaceous limestone, Upper Triassic dolomite, and Jurassic limestones and dolomites (Čar and Šebela, 2001), providing a unique example in Slovene caves. The main cave entrances are formed at the tectonic contact between Eocene flysch rocks and Upper Cretaceous limestones (Čar and Šebela, 2001) at the sinking point of the River Lokva at 464 m (confirmed with GNSS receiver). Predjama Cave is developed at more levels with a medieval castle in front of the cave shelter (Fig. 1.6).

The main aim of this monograph is to present and analyze the results of microclimatic data sets from show caves in Slovenia, mostly from Postojna and Predjama Caves, and to define significant changes and their causes during the 10-year monitoring period. Understanding cave climates and cave biology in show caves is between top priorities for realization of sustainable use of the caves for tourism.

In Slovenia, we have a law protecting underground caves (Zakon o varstvu podzemnih jam—ZVPJ), which came into effect in 2004 (http://www.pisrs.si/Pis.web/pregledPredpisa?id=ZAKO2068). The

Figure 1.6 Predjama Castle is located in a limestone wall in front of the cave shelter of the upper Predjama Cave. *(Photo: S. Šebela.)*

12th article of the ZVPJ law determines monitoring of the state of the caves, and includes the observation and monitoring of the known caves in the Republic of Slovenia, their living world and cave inventory, ecological conditions in caves, pollution in caves, and the number of visitors. The 13th article of the ZVPJ law prescribes that suggestions of measures needed for preservation and remediation of caves, especially operational program for cleaning of polluted caves, must be included in the regular reports on the state of caves.

Show caves in Slovenia need a cave guardian (26th article of ZVPJ law). All karst caves in Slovenia are natural treasures of national importance and are natural public goods that belong to the State. Also, cave infrastructure is a property of the State (5th article of ZVPJ law). Postojna Cave, which is the most visited show cave in Slovenia, is managed by a private company with a concession contract. Predjama Cave is also managed by the same company.

This monographic work provides extensive analysis of the microclimate data obtained from field measurements in the period 2009–20, and is especially oriented to the reciprocity between natural and anthropogenic impacts on cave climates in the show caves in Slovenia. The aim of the research was for examples from microclimate studies from Slovene show caves to provide information that would be valuable for similar show caves worldwide to strengthen international scientific cooperation and understand new scientific challenges.

This book is written on the occasions of:

2021—International Year of Caves and Karst;

1991—2021—30 years of the Republic of Slovenia;

1818—2021—203rd anniversary of the discovery of the most extensive section of Postojna Cave;

1819—2021—202nd anniversary of modern tourism in Postojna Cave;

1819—2021—202nd anniversary of modern tourism in Škocjan Caves;

1947—2021—74th anniversary of the Karst Research Institute ZRC SAZU;

1981—2021—40th anniversary of the ZRC SAZU.

CHAPTER 2

Purpose of research

Microclimatic monitoring presented in this monograph started in 2009 within expert control for the sustainable use of the Postojna and Predjama Caves for tourism. Some scientific issues related to this subject have already been published (Šebela and Turk, 2011a, b, Šebela et al., 2013, Šebela and Turk 2014a, Gregorič et al., 2014, Šebela et al., 2015a, b, Šebela and Turk 2018, Prelovšek et al., 2018, Pipan et al., 2019), however a lot of unprocessed data remain, which have not been fully analyzed yet and it is hoped that with this monograph the gap will narrow. In additionally to Postojna and Predjama Caves, some preliminary microclimatic results for other Slovene show caves such as Škocjan Caves (inscribed in UNESCO natural heritage), Županova Jama, and Kostanjeviška Jama are included also.

The purpose of the research described in this monograph is to present the complexity of the microclimatic research in selected show caves in Slovenia, especially regarding cave air temperatures and carbon dioxide concentrations. The cave CO_2 levels and CO_2 exchange take part in processes such as calcite speleothem growth (Faimon et al., 2020). Exchange of CO_2 between the underground caves and the exterior is a significant part of the carbon cycle and should be involved in carbon balances (Faimon et al., 2020).

Maximum perturbation cave temperature depends on the number of visitors and the duration of their stay in the cave: the higher the number of visitors and the smaller the cave, the greater is the temperature increase (Rakovec, 2020). When visitors leave the cave, only the exchange with walls remains and the temperature starts to recover toward natural conditions.

Essential increases happen in smaller caves with many visitors. Recovery back to natural conditions is quicker in smaller caves, and slower in larger ones (Rakovec, 2020).

Show caves in Slovenia are important for economic income, tourist experiences, employment, recreation, and also for scientific research. Microclimatic monitoring in show caves is organized as part of environmental protection monitoring. There are caves that have been show caves for 200 years (Šebela et al., 2015a, b) and we are asking ourselves what the

Natural and Anthropogenic Impacts on Cave Climates
ISBN 978-0-12-822954-5
https://doi.org/10.1016/B978-0-12-822954-5.00002-0

impact of long-term visits to a fragile cave environment is and what can be improved to preserve sustainable use of show caves for tourism in the future.

In this work we use the terms "cave meteorology" and "cave climate" (or "cave microclimate" when we speak of smaller portions of the cave system). Badino (2010) used both terms, where "climatic" was used to indicate average values of a condition, while "meteorological" referred to fluctuations around it and transient phenomena. In our caves, 10-year air temperature time series with mean annual temperature values are treated as the "cave climate," while "cave meteorology" is related to shorter, selected periods with descriptions of interactions between different meteorological parameters such as cave air temperature, carbon dioxide concentrations, direction of air circulation, precipitation, and external air temperature.

The first impressions gained about cave air temperatures can be very disappointing, because air temperature oscillations over 1 year can be less than 1°C, which when compared to outside air temperature can be described as "constant temperature," pointing to a very stable underground environment. However, in 1965 (Cropley, 1965) it was confirmed that cave temperatures are not necessarily constant but may vary widely. This was confirmed also in our caves with continuous air temperature monitoring over an extended number of years, which showed interesting and sometimes unexpected temperature differences between years. Over the last 10 years, winters in Slovenia have been mild meaning that cold outside air did not enter Postojna Cave in the way that it would in the case of very cold winters. An exception was seen in February 2012 with more days with a long northern "bora" wind [see Chapter 5.1.3.3, Lepe Jame (Postojna 3 site)], which significantly lowered air temperature deep inside the cave. In 2010, the area of Postojna witnessed the highest floods in the previous 50 (or even 100) years, which impacted the water temperature in the rimstone pool [see Chapter 5.1.3.3, Lepe Jame (Postojna 3 site)]. In an unplanned situation, a temperature instrument was unintentionally placed on the cave sediments, causing the air temperature at the ground to be measured instead of the cave air (see Chapter 5.1.3.5, Koncertna Dvorana). The results were interesting and showed for how much the cold outside air can lower the cave air (and/or ground floor cave sediments) temperature in winter.

The oldest meteorological measurements in Slovene caves (Schmidl, 1854; Vercelli, 1931; Crestani and Anelli, 1939, Anelli, 1941—44) provide remarkable data despite the fact that the precision of measuring instruments was not as good as nowadays. Comparing historic cave air temperatures with recent measurements (Pipan et al., 2019) gives significant information about temperature changes (Domínguez-Villar et al., 2015). In Postojna Cave, historic air temperature measurements were compared with recent measurements (see Chapter 5.1.3.2, Velika Gora, and Chapter 5.1.3.6, Sepolcro).

The studied caves are large cave systems where microclimatic monitoring at more locations with the use of modern good-quality instruments that can store huge data sets and are suitable for measurements in extreme environments over many years is needed. The position of measurement equipment in a show cave is important because changes can appear if the same cave profile is measured at the bottom of the passage compared with just below the cave ceiling. To get representative long-term data sets it is the best not to change the instrument position, but rather to add new monitoring devices for a new monitoring locations; this is what our microclimatic monitoring experiences have shown to be the best solution for our caves.

Microclimatic monitoring locations are placed within the most heavily visited passages and almost unvisited passages to see the differences in microclimate parameters in touristic and non-touristic parts of the same cave system. The aim of the research was to establish a confident data base of measured parameters and to perform their research analyses. One of the main issues was to determine the factors that impact cave climates. Are those natural factors, such as the impact of outside climates on cave climates, or anthropogenic impacts, such as the transfer of heat and carbon dioxide by visitors? The fact is that both causes interact and it is not easy to distinguish between them. In this sense, the microclimate of the heavily visited Postojna Cave was compared with that of the less visited Predjama Cave where the anthropogenic impact due to visitors was negligible.

Based on microclimates, more long-term statistical analyses have been derived and the meteorological characteristics of different parts of the cave systems have been described. The aim of the study was to suggest guidelines for sustainable use of the studied show caves where possible. Guidelines and suggestions can be used by cave managers, park services, and agencies and ministries responsible for nature conservation. Where guidelines were not clearly determined, the microclimate parameters were documented as the state-of-the-art, for use in future studies into the development of cave microclimates.

In recent years climate studies have increased because of the worldwide issue of climatic change (Kalnay et al., 1996; Parmesan and Yohe, 2003, etc.). Also, cave paleoclimate (Baldini et al., 2002; White, 2007, etc.) and cave climate studies are growing (de Freitas, 2010; Badino, 2010, etc.). The characteristics of cave sediments are applied in paleoclimate archives (White, 2007). There are potential applications of different dating methods for cave sediments to late Holocene climates with implications for the evaluation of current concerns over global warming (White, 2007).

In accordance with the expected gradual warming throughout Europe, Slovenia will also be subject to a significant temperature rise, ranging from 1°C to 4°C in annual mean temperature by 2100 (http://meteo.arso.gov.si/uploads/probase/www/climate/text/en/publications/OPS21_brosura_ENG.pdf). The mean annual air temperature increase in the capital of Slovenia (Ljubljana) was +1.1°C over the last 100 years (Ogrin, 2015). One of the research questions asked was "What can we expect in show caves regarding outside heat transmission to the cave environment?"

In this work, the goal was to analyze the relation between natural and anthropogenic reasons for changes to cave climates in show caves. Is the rise in cave air temperature due to heating of the outside atmosphere a natural or anthropogenic cause, or possible both? The fact is that global climate changes have a strong impact on the underground environment also.

The results of microclimate analyses in caves presented in this book are based on real, field, "in situ" observations. Explications of particular situations related to specific monitoring locations are described on the bases of theoretical facts of underground meteorology, mostly from Smithson (1991) and Badino (1995, 1999, 2010).

In the first part of the book (see Chapter 3, Similar Studies Worldwide) more theoretical outlines with some case studies, that strongly impacted my view of cave microclimate in Slovene show caves, are presented. The subchapters of Chapter 5 include descriptions of individual Postojna and Predjama Cave microclimate monitoring sites, while Chapter 5 (Microclimatic Monitoring in Show Caves) is not limited only to Postojna and Predjama Caves, but includes also some preliminary research regarding air temperature monitoring in other Slovene show caves. Chapter 6 (Sustainable Use of Show Caves) discusses the most important issues that need to be taken into consideration to successfully fulfill the aim of sustainable use of Postojna and Predjama Caves for tourism.

Examples from Slovene show caves can be useful for other similar show caves around the world. It is important to be regularly updated with the latest trends in management implications in show caves regarding different environmental policies. It is valuable also to learn from good practices, from know-how practices, and to apply such solutions in show caves, for their benefit. The international exchange of good practices is important also for improvement of the management of the show caves and for international scientific cooperation.

CHAPTER 3

Worldwide studies of cave meteorology and cave climate

There is no way to include all publications related to cave climates. This is why this chapters outlines the major publications which have been the most important for interpretations of our results obtained from the case studies presented in this monograph.

The first author to greatly impact my view of cave meteorology and cave climate was the late Giovanni Badino (1953–2017). His works, such as *Physics of the Subsurface Climate* represent the basic literature (Badino, 1995, 1999) for those interested in cave meteorological processes.

According to Badino (1999) the behavior of the air that fills underground voids is fundamental to understanding the global structure of the inner karst massif. The air circulation inside caves is a small part of the greater atmospheric air circulation, what means they share the same air, the same physics, and basic phenomena (Badino, 1999). The climatology of caves includes complex exchanges of energy and matter between fluids in a thermically quasi-closed underground system (Badino, 1999).

Slovene caves are very humid. In caves, the humidity is close to saturation. Badino (1999) described that the behavior of humid air is different to that of dry air. The gases that expand in the state of relative thermic isolation become colder. The dry air becomes colder by 9.7°C per kilometer in vertical distance, while humid air cools by 5–6°C per vertical kilometer. The processes of air mass transfer in the vertical direction are very important (Badino, 1999). The thermic gradient in a cave system does not depend greatly on the mean temperature values of the cave (Badino, 1999).

For caves, it is typical that the air temperature shows only slight seasonal changes, which can be detected only with precise instruments. Caves feel cold in summer and warm in winter, and the cave air temperature is the most similar to the temperature of the major mass of the karst system (Badino, 1999).

The heat can be stored in the air, water, and/or in the karst rock. Rock is also an excellent thermic isolator. Badino (1999) calculated that if we additionally heat the air entering an Alpine cave system from 10 to 20°C for

Natural and Anthropogenic Impacts on Cave Climates
ISBN 978-0-12-822954-5
https://doi.org/10.1016/B978-0-12-822954-5.00003-2

1000 s an air temperature of 10.003°C is obtained, and if we heat it for 3.8 months we get 10.3°C. He wanted to show that even the warmest seasons do not greatly affect the underground cave temperatures in Alpine caves (Badino, 1999).

It is generally accepted that air temperature is slightly higher than water temperature, which means between these fluids there is a 1° temperature difference (Badino, 1999). The cave air temperature can be higher than the water air temperature only in the case that the karst system is not isolated and there is an input of heat from the exterior (Badino, 1999).

Another important factor in cave meteorology is the air circulation. It is caused by differences in air pressure: the air arrives from where the pressure is higher and goes to the place where it is lower (Badino, 1999). Air circulation in caves depends upon the structure of the passages and can provide important speleological information (Badino, 1999). Cave air circulation depends also on the temperature difference between the cave and exterior. Cave air circulation is apparent when the external temperature is different to that in the cave by more than ΔT (Badino, 1999). This last situation is described from air temperature monitoring sites in Postojna and Predjama Caves.

In caves, convective circulation arises. The cold air that exits at lower cave entrances pushes the same volume of external air higher, and to do this it has to expend energy. In general, caves in winter blow air out of higher cave entrances and take in air at lower entrances; in summer it is the reverse (Badino, 1999).

Cave temperature is generally strictly connected to the external climate, and global climatic changes can influence the cave climate (Badino, 2004), as is discussed in Chapter 5.1.7 (Anthropogenic and natural impacts on Postojna Cave microclimate), Chapter 5.2.4 (Natural impacts on Predjama Cave microclimate), and Chapter 6 (Sustainable use of show caves). It is interesting to note that in 1999 Badino showed Alpine caves as very stable temperature environments with an insignificant impact of external heating, but in 2004 he published his work about *Global Climatic Change and Cave Temperatures* (Badino, 2004).

Based on meteorological measurements in an Alpine cave, it was estimated that the explored volume of the underground system was less than half its probable real extension (Badino and Chignola, 2019). Meteorological measurements in caves have been shown to be important for speleological explorations for possible new passages.

The use of original experimental data and experimental study of underground meteorology are difficult because they are time-dependent and have very small amplitude, however measurements show that their variability from one cave to another and from point to point in the same cave is surprisingly high (Badino, 2010). The same was confirmed in Postojna and Predjama Caves.

According to Badino (2010), cave physics studies are important to protect caves, especially show caves, against anthropogenic impacts. Despite the fact that the underground temperature increases with depth, caves are usually very cold and unaffected by geothermal heat flux (except in geothermically active areas). The reason for this is the deep vadose zone, with percolating meteoric water, which equilibrates the rock mass with external temperature (Badino, 2010). It is assumed that a cave temperature is equal to the temperature of the surrounding rock, which has the temperature of infiltrating water, which is almost equal to the local annual mean temperature (Badino, 2010). The estimation of the outside mean air temperature is of extreme importance for understanding cave physics (Badino, 2010).

The cave air flux does not depend only on the external atmosphere but mainly depends on the cave morphology: entrances, shape, internal impedance, vertical passages, narrow passages, etc. (Badino, 2010). The cave morphology can affect the cave air temperature, creating cold and warm air traps, through a process that can be included in the "fluid selection" and can cause strong thermal stratification (Badino, 2010). A hermetically closed cave passage, developing upward (downward) from the main conduit, can trap the air flowing through it in an especially warmer (colder) season, thus creating air bubbles, which remain untouched when the airflows become colder (warmer) than usual (Badino, 2010). This trapping effect is described also in Postojna Cave [see Chapter 5.1.4, Črna and Pivka Jama, Chapter 5.1.3.2, Velika Gora, Section 5.1.3.4, Lepe Jame (Postojna 2 site) and Chapter 5.1.5, Otoška Jama]. Even if water and air are transported through caves, it is the water which usually plays the major role in cave temperatures (Badino, 2010). In the upper part of a karst conduit, the air is often warmer than the air at the bottom, and the air temperature changes with height (temperature sedimentation). We note that this always implies energy exchanges, instabilities, and that the karst system is not thermodynamically closed, but in contact with some time-dependent external processes (Badino, 2010).

The lapse rate in a cave atmosphere is significantly lower (in absolute value) than the external one, and these experimental underground lapse rates are not adiabatic (Badino, 2010). Water and air move vertically, and their temperature changes, but at different rates: this creates temperature disequilibria inside the karst system, therefore we have continuous energy exchange between air, water, and rock (Badino, 2010).

Additionally, it is important to understand what happens when the vapor pressure in caves is slightly below or above the equilibrium level (under- and supersaturated), because in such cases we refer to evaporation and condensation that are very important for the speleogenetic development of cave systems (Badino, 2010).

Air circulation in caves can be related to convective air circulation or barometric circulation (Badino, 2010). Convective air circulation in caves depends on the buoyancy and friction of the cave air, and air flow velocities would depend on the square root of ΔT, which is the difference between the instantaneous cave and external temperatures (Badino, 2010). In the case of the studied locations at Postojna and Predjama Caves the convective air circulation is relevant.

The observations of cave meteorology show relatively small amplitude variations for cave temperature and humidity, compared to the external variations, however, this does not justify the common thought that, in a cave, either climatic or meteorological processes do not exist (Badino, 2010). This observation was described by Badino (2010) is a typical statement for Postojna and Predjama Caves.

Interrelationships between cave and outside air temperatures have been described by Smithson (1991) for Poole's Cavern in the Peak District of Derbyshire (Great Britain). This work was very important for my view of underground meteorology because we understood that we need to establish a local surface air temperature monitoring point above the Postojna Cave passages to get the best comparison between the underground passages and surface conditions. In fact, the surface temperature monitoring location at hourly intervals was stabilized in August 2009 (see Chapter 4, Methods) in the forest above the Postojna Cave.

Each cave system has an element of uniqueness while following the general principles of the cave temperature models (Smithson, 1991). Cave climate is a function of cave air exchange with the outside atmosphere. Cave airflow can take place through gravity, density-induced convection, and forced advection (Smithson, 1991). In the Roman Chamber, which is located close to the cave entrance of Poole's Cavern, the most stable situation occurred during the summer and early autumn, when outside temperatures were well above those of the cave (Smithson, 1991). When outside temperatures fall below cave temperatures, the usual density-induced outflow of cold air from the cave is reversed, meaning that colder air flows in the cave along its lowest part (Smithson, 1991). Such situations are typical for Postojna and Predjama Caves, pointing to winter and summer ventilation regimes in caves. Smithson (1991) also described that caves with openings in especially west and southwest directions are more likely to be affected by advectional interchange.

Cropley (1965) described the occurrence of peak cave temperature lagging behind the corresponding peak in surface temperature. The lag increases by about 1 week per 0.3 m of depth (Cropley 1965). Such temperature lags are described in Postojna Cave also [see Chapter 5.1.3.3, Lepe Jame (Postojna 3 site), Chapter 5.1.3.2, Velika Gora].

Among recent publications I need to mention a good contribution about cave air temperature as a key parameter for climatic processes by Killing-Heinze et al. (2017). These authors identified thermal- or convective-driven influences, and also some clear barometric effects on cave climate research in Carlsbad Caverns (United States) (Killing-Heinze et al., 2017). Because of the complexity of the Carlsbad Caverns some parts are strongly influenced by outside weather conditions, while others show more long-term signals (Killing-Heinze et al., 2017). To ensure a balance between visitors and sustainable use of the cave it is very important to understand the climatic processes in a show cave (Killing-Heinze et al., 2017).

The most important for the Carlsbad Caverns climate is the convective system of airflow (Killing-Heinze et al., 2017). During winter, cold air flows into the Big Room replacing warmer and moister air and causing a cold air trap (Killing-Heinze et al., 2017), as is also the case in Črna Jama (see Chapter 5.1.4, Črna and Pivka Jama). At Devil's Spring (a pool near the natural entrance to Carlsbad Caverns) during the winter months cold air flows near the ground into the cave replacing warmer air (see Chapter 5.1.1, Postojna Cave entrance part) that then forms a convective compensation flow along the ceiling and out of the cave. However, during the summer, when rising temperature leads to reduced air exchange between the cave and the surface, the climate at Devil's Spring is independent of variations in the outside weather (Killing-Heinze et al., 2017). There are particular phenomena during midday when the air temperatures at the ground and ceiling tend to converge. This happens when high solar radiation leads to turbulence in the outside air that affects the cave air also (Killing-Heinze et al., 2017).

In the narrow part of the Left Hand Tunnel inside Carlsbad Caverns, a graph of the air temperature recorded every 5 min shows a double wave with two peaks per day. This indicates a ventilation system that is not convective as described at Devil's Spring but can only be caused by barometric events (Killing-Heinze et al., 2017). Long-lasting barometric ventilation systems develop mainly in parts of the cave where the large ratio between their volume and the diameter of their entrances inhibits fast pressure equalization (Killing-Heinze et al., 2017).

Interesting results were obtained from October 1 to 16, 2013, when Carlsbad Caverns were temporarily closed to visitors. It was shown that the air temperature heated up several times per day during guided tours (Killing-Heinze et al., 2017). It was detected also that Carlsbad Caverns have constantly higher temperature year-round (Killing-Heinze et al., 2017). Research confirmed that the air inside the Big Room and King's Palace tour is constantly warmer than it would have been without the development of Carlsbad Caverns for tourism (Killing-Heinze et al., 2017). It was additionally shown that in the Big Room and King's Palace, as the most visited cave sections, the convective ventilation, as well as the influence of the barometric ventilation, are almost completely overwhelmed by the anthropogenic influence (Killing-Heinze et al., 2017).

In the study by Killing-Heinze et al. (2017) there is an interesting statement that it remains unsolved how anthropogenic modifications of the cave climate at Carlsbad Caverns affect cave fauna and speleothems. This is one of the most important issues in show caves and is applicable also in the greatly visited Postojna Cave.

Complex research of long-term climate monitoring over a number of years in four French prehistoric caves was presented in 2014 by Bourges et al. Some prehistoric painted caves suffered irreversible degradations as a consequence of misperception of conservation issues and subsequent mismanagement (Bourges et al., 2014). Many prehistoric caves have suffered from degradation due to tourist visits and archeological operations, and, in general, due to anthropogenic and natural effects (Bourges et al., 2014).

The studied caves were divided into: confined state, intermediate (or semiconfined), and open-state caves (Bourges et al., 2014). A single-opening cave in winter has an open regime because cold and dense outside air is pulled in the cave by thermo-convection (Bourges et al., 2014). In the same cave in summer, stable stratification arises between the warmer outside air and colder cave air, causing the cave to be in a stabilized confined state (Bourges et al., 2014).

In our case, such an example is Črna Jama (see Chapter 5.1.4, Črna and Pivka Jama). Caves with two openings at different levels present an unconfined state in summer and winter (Bourges et al., 2014). Confined cave rooms have very low variations in their thermodynamic parameters (temperature changes of less than a few 0.01°C) and ensure stable conservation of their natural physical systems, while open cave areas (such as those close to cave entrances) have unstable parameters and can represent more difficult issues regarding cave conservation (Bourges et al., 2014).

Marsoulas prehistoric caves are in a non-confined state that is evidenced by a lack of any correlation between cave temperature and pressure first time derivated (dP/dt) (Bourges et al., 2014). In the example of Pech Merle Cave, the progressive transition of the cave atmosphere from an open state in the cold season to a permanent confined state can be attributed to the improvement of the door closure (Bourges et al., 2014). Most painted caves have been disturbed by enlarging existing narrow passages or by opening new passages, and also by visitors (Bourges et al., 2014). This is why monitoring of cave air temperature is recommended to control sudden changes in cave climates (Bourges et al., 2014).

In the case of the famous prehistoric Lascaux Cave in France, carbon dioxide concentrations were measured to be 6200 ppm in summer and 400 ppm in winter (Peyraube et al., 2018). Minimum and maximum carbon dioxide levels can affect the state of the cave system, as speleothem growth on the cave walls with prehistoric paintings and engravings can be affected by calcite dissolution or precipitation (Peyraube et al., 2018). Lascaux Cave was opened for visitors in the period 1940–63. The cave is now accessible only for scientists for a restricted period, because its conservation is of the greatest importance (Peyraube et al., 2018).

A model of heat and mass transfer in a shallow cave (such as Lascaux Cave) included conductive transfer in the karst massif, convective and radiative transfer in the cave, and evaporation/condensation on the cave walls (Guerrier et al., 2019). Such a calculated model can improve the management of caves, especially those with Paleolithic paintings because it can identify areas which are most exposed to weathering (Guerrier et al., 2019).

Convection is the principal mechanism for the transport of matter within the gaseous mass of Altamira Prehistoric Cave in Spain (Villar et al., 1984). Two main causes for the deterioration processes that may attack the paintings on cave walls are evaporation and condensation of water on the rock surface (Villar et al., 1984).

Altamira Cave shows two distinctive ventilation modes. From October to May, as a result of reduced exchange with the outside atmosphere, the cave is a CO_2 reservoir (recharge stage), while in the period July to September it is in a discharge state, actively exchanging gases with the outside (Gázquez et al., 2016). Experiments that evaluated the impact of visitors on the $\delta^{13}C-CO_2$ showed significant changes in this parameter caused by the CO_2 exhaled by visitors, which has a distinctive isotopic signature (Gázquez et al., 2016). The isotopic composition $\delta^{13}C-CO_2$ in the cave atmosphere can be utilized together with other traditional parameters (air temperature, relative humidity, etc.) to evaluate the tourist impact on prehistoric and show caves (Gázquez et al., 2016).

The prehistoric Candamo Cave (Spain) with rock art has been seriously changed due to mass tourism (Hoyos et al., 1998). Microclimatic monitoring (CO_2, air temperature, humidity, ^{222}Rn) was established during the period when the cave was closed to the public and during an experimental period of controlled visits (Hoyos et al., 1998). It was found that the summer period (July–October) was the most favorable time for visits because of the presence of the greatest ventilation and thus, the greatest capacity for recovery (Hoyos et al., 1998). In this article it was stressed that calculation of visitor capacity for a cave must be based on environmental parameters but extensive multidisciplinary knowledge of the problem cannot be excluded (Hoyos et al., 1998). Visitor capacity was calculated as 29 visitors per day (Hoyos et al., 1998).

Very detailed cave climate studies have been performed in the karst of Czechia (Faimon et al., 2006, Faimon et al., 2012, Lang et al., 2017a, b, Faimon et al., 2020, etc.). In the case of Císařská Cave (Moravian Karst, Czech Republic) three ventilation modes, based on intensive exchange between cave air and the exterior, were detected: UAF mode (upward airflow), DAF mode (downward airflow), and transitional mode (Faimon et al., 2012).

Cave air flows are mostly related to the temperature difference between outside and cave atmospheres, with two ventilation regimes (Lang et al., 2017a, b). When $\Delta T < 0$ ($\Delta T = T_{out} - T_{cave}$), upward air flow (UAF mode) occurs and when $\Delta T > 0$, downward flow air ventilation (DAF mode) is taking place (Faimon et al., 2012, Lang et al., 2017a, b). The ventilation modes of Faimon et al. (2012) were determined and explained also in our caves (see Chapter 5.2.4 Natural impacts on Predjama Cave microclimate).

In the same show cave (Císařská Cave, Moravian Karst, Czech Republic) a study related to CO_2 fluxes was performed to contribute to better preservation of the cave environment (Faimon et al., 2006). Two types of CO_2 fluxes into the cave were determined: the natural input and an anthropogenic input (Faimon et al., 2006). Natural CO_2 input was determined as CO_2 degassing of the drip waters. Anthropogenic CO_2 input is related to visitors. This input can be added to the natural CO_2 values, especially if the period of visitors staying in the cave had been slightly extended (from 2.85 to 4 h). In such a case, the drip water CO_2 degassing would be inverted into CO_2 dissolution and drip water supersaturation would decrease (Faimon et al., 2006). It was calculated that the simultaneous presence of 100 persons in the studied cave chamber for 14 h could cause the situation that drip water can become aggressive to calcite speleothemes in the cave (Faimon et al., 2006). Similarly to Císarská Cave, natural and anthropogenic CO_2 values have been described in Postojna Cave also (see Chapter 5.1.2, Biospeleološka Postaja).

The field CO_2 measurements performed in Comblain-au-Pont Cave (Dinant Karst Basin, Belgium) and in Moravian karst caves, together with published data about CO_2 levels in caves worldwide have been studied by Lang et al. (2017a). In caves there is direct diffusive CO_2 flux from soils/epikarst, the indirect CO_2 flux derived from drip water degassing, and the input/output CO_2 fluxes linked to cave ventilation (Lang et al., 2017a). A dynamic CO_2 model indicated that the additional CO_2 flux could be convective advective CO_2 flux from soil/epikarst driven by airflow (cave ventilation) and enhanced soil/epikarst CO_2 concentrations (Lang et al., 2017a). Maximum CO_2 levels of up to 70,000 ppmv have been found in caves worldwide (Lang et al., 2017a).

Understanding temporary anthropogenic phenomena impacting cave microclimates is important for preparation of the control of visiting tours to show caves (Lang et al., 2017b). This was especially challenging in the case of Balcarka Cave (Moravian Karst, Czech Republic), where the show cave is equipped with doors which open frequently during visits (Lang et al., 2017b). The impact of door opening on cave ventilation and CO_2 was studied in this cave (Lang et al., 2017b).

Winter CO_2 concentrations at Balcarka Cave are similar to the external values and represent natural CO_2 levels because the cave is closed in winter (Lang et al., 2017b). In summer, when the cave is opened for visitors, CO_2 can reach 7000 ppmv (Lang et al., 2017b). Balcarka Cave is a dynamic cave system significantly influenced by cave ventilation (Lang et al., 2017b).

It has an open lower entrance and upper hidden openings that allow the air and CO_2 to penetrate into the cave through soils and epikarst (Lang et al., 2017b). Whereas the lower CO_2 concentrations are associated with UAF ventilation mode, higher CO_2 concentrations are associated with DAF mode due to direct advective CO_2 fluxes from soil and epikarst (Lang et al., 2017b). This study showed that even short-term changes in cave ventilation due to door opening may significantly influence cave microclimate (Lang et al., 2017b). In Škocjan Caves in Slovenia we have an artificial tunnel closed with doors that lead to natural cave passages. Also, there is a strong correlation between door closing and door opening due to visits shown in CO_2 concentrations (Rakovec, 2020) due to forced ventilation.

Intrusion of the air from outside always essentially decreases the CO_2 concentration. This occurs when an entrance to the cave is opened for visitors and some of the outer air can intrude into the cave (Rakovec, 2020). Once the door is closed, the air will settle down again. For the air as a whole in such a cave, initially a rather fast eddy diffusion dominates, and then the exchange with walls is the only process that leads to slower natural conditions (Rakovec, 2020).

The exchange of gases between the breathing spot in Hranice Karst (Czech Republic) and underground caves was studied by Faimon et al. (2020). They distinguished between breathing spots and blow-holes (breathing openings). A blow-hole simply facilitates air exchange, while a breathing spot includes exchanged air with additional CO_2 from the overlying soil (Faimon et al., 2020). It was shown that warming of the soil is a dominant process in the UAF mode (upward air flow), and CO_2 advective flux into the cave is the main process in the DAF mode (downward air flow) (Faimon et al., 2020). Exchange of air currents between surface and underground passages through "unknown passages" is significant for Postojna Cave also [see Chapter 5.1.3.4, Lepe Jame (Postojna 2 site)].

Knowledge of heat fluxes and temperature distribution in continental rocks is of great importance for many engineering aspects (tunneling, mining, geothermal research, etc.) (Lütscher and Jeannin, 2004). In temperate karst regions, water circulation is usually high enough to completely "drain-out" the geothermal heat flux at the bottom of karst systems (phreatic zone) (Lütscher and Jeannin, 2004). Air circulation can greatly dominate water infiltration in a karst vadose zone, consequently, temperature gradients within a vadose zone are similar to the lapse rate of humid air ($\sim$0.5°C per 100 m) (Lütscher and Jeannin, 2004).

A reduction of the ice in caves in the Jura Mountains and its association with climate change in glaciers was described by Lütscher et al. (2005). Ice caves have been described in many parts of the world at altitudes where the mean annual air temperature is several degrees above 0°C (Lütscher et al., 2005). Recent cave ice fluctuations show a negative trend in ice mass balance, with an increasing trend since 1980. Because most of these ice caves act as cold air traps, this negative mass balance is mainly attributed to higher winter temperatures and reduced snow precipitation at low altitudes (Lütscher et al., 2005). The importance of winter temperature and precipitation regimes for the conservation of perennial cave ice has been outlined (Lütscher et al. 2005). Indications of former ice fillings that have been found in different caves of the Jura Mountains outline their potential role as paleoclimatic markers (Lütscher et al., 2005). An example of a typical cold air trap cave in the Postojna Cave system is Črna Jama (see Chapter 5.1.4, Črna and Pivka Jama).

Nerja Cave in Malaga province in Spain is a famous well-visited show cave, where numerous studies relating the cave microclimate and environmental protection issues have been published (Carrasco et al., 2002; Liñán et al., 2008; Liñán & del Rosal, 2015; Liñán et al., 2018; Liñán et al., 2020, etc.).

In 2002, Nerja Cave received more than 500,000 visitors per year and since 1993 underground environmental monitoring has been carried out (Carrasco et al., 2002). The continuous record of physical and chemical parameters of drip water, its daily outflow, air temperature and relative humidity, CO_2 concentrations, and rock temperature show the human influence (Carrasco et al., 2002). Some environmental parameters are altered due to anthropogenic impact, such as: cave air temperature increases by 0.2°C for each 1000 visitors/day, a daily increase in relative air humidity is between 2% and 3%, CO_2 concentrations in air increase up to 500 and 700 ppm with low visit numbers and up to 2800 ppm during high-visitor periods, the temperature of the rock rises between 0.02°C and 0.15°C per day, and the P_{CO2} of drip water increases during a big influx of visits and the saturation index of carbonated minerals decreases (Carrasco et al., 2002).

The temporal variation of CO_2 concentrations in the air of the Nerja Cave is strongly influenced by natural ventilation, which is determined by the difference between external and cave air temperatures (Liñán et al., 2008). Additionally, a positive correlation between CO_2 concentrations and visitor numbers was observed during August, the month that receives the most visits throughout the year, averaging 100,000 (Liñán et al., 2008).

A natural convective air flow, which follows a seasonal model common in caves, known as "chimney," characterized by at least two entrances at different altitudes, is present at Nerja Cave (Liñán & del Rosal, 2015). Regarding the model of air flow for known entrances of the Nerja Cave, it was shown that there must be some connection with Pintada Cave that is in the vicinity. A connection inaccessible to humans between both caves was described with a general model of air flow circulation between them, which allows the removal of anthropogenic impact in the Nerja Cave during the most visited season in the year (Liñán & del Rosal 2015).

Based on microclimate and gravimetric data in the area of the Nerja-Pintada cave system (Spain), the existence of an unknown great cavity located close to the Nerja Cave and with a direct influence on its ventilation was determined (Liñán et al., 2020). It was confirmed that analysis of the data obtained from additional research methods (as gravimetry) together with microclimatic monitoring is a useful tool in the characterization of karst systems (Liñán et al., 2020).

Understanding of microclimatic processes is especially important in the management of show caves (de Freitas, 2010). Unlike microclimates in the atmosphere–land boundary layer, which are characterized by vertical exchanges, processes determining climate in all but almost closed caves are dominated by the advection of heat and moisture (de Freitas, 2010). Thermodynamic aspects of external air–cave air interactions are important to explain spatial, short-term, and seasonal variations of thermal and moisture states of the cave atmosphere (de Freitas, 2010). De Freitas (2010) argued that cave management is not simply a matter of determining usage levels or the carrying capacity of caves but also involves determining environmental management techniques that are appropriate to a particular cave condition or environmental state that should prevail. I have found this very useful also in the case of Postojna Cave. Further, De Freitas (2010) wrote that the impact of visitors on cave climate is not linear as, for example, the effect of a group of 15 people may be more than three times the impact of a group of five.

The Waitomo Glowworm Cave in New Zealand, with a large number of visitors and the presence of cave fauna, is a significant case study of sustainable management because it has been continuously used as a tourist cave for over 120 years (De Freitas, 2010). Monitoring of microclimatic indicators at Waitomo Cave includes air flow, air temperature and humidity, carbon dioxide, radon, and condensation (De Freitas, 2010).

In the best known show cave in Australia—Jenolan Caves—abrupt CO_2 increases were usually short-lived, but when tours were frequent, the CO_2 concentrations did not decrease to the pre-tour levels before the next tour (Baker, 2014). Simultaneous spikes in air temperature and CO_2 corresponded with tours through the cave (Baker, 2014). CO_2 was determined to be a crucial factor in many of the processes that form caves and their speleothems (Baker, 2014).

Of numerous US caves we need to mention Jewel and Wind Caves, which are large barometric cave systems in South Dakota (Pflitsch et al., 2010). In barometric caves pressure variations of the atmosphere enforce a temperature-independent compensating air flow (Palmer, 2007; Pflitsch et al., 2010), while in a thermal cave density differences of unequally tempered air masses lead to compensating air flows (Pflitsch et al., 2010). Barometric caves show smaller differences between summer and winter than thermal caves; this effect is due to the seasonal variability and stability of passing pressure systems (Pflitsch et al., 2010). There are also evidences that thermal and barometric effects can exist inside a single cave system (Boes et al., 1997). For Postojna and Predjama Caves we did not find reliable proof for a barometric cave, but this does not mean that there are no side remote passages not included in microclimatic monitoring, that could also have characteristics of barometric caves also.

Finally, we can only repeat, after Cigna and Forti (2013), that it is necessary to constantly control at least the most sensitive cave parameters in order to immediately correct cave management as soon as the first adverse effects appear and avoid seriously damage to the show cave.

CHAPTER 4

Methods

The most important measured parameters of microclimatic monitoring in Postojna and Predjama Caves are air temperature values and carbon dioxide concentrations. To better understand specific conditions, measurements of the direction and velocity of air currents and air pressure have been added for selected time periods. To compare the cave microclimate situation with outside cave conditions, external air temperature measurements have been organized parallel to cave monitoring sites. Precipitation data obtained by the Ministry of the Environment and Spatial Planning, Slovenian Environmental Agency (https://meteo.arso.gov.si/met/sl/archive/), have been used for selected time periods.

Air temperature was measured at hourly intervals with Eijkelkamp barodivers. The accuracy of the measurements was ±0.1°C with resolution ±0.01°C. At Postojna Cave entrance monitoring site (see Chapter 5.1.1, Postojna Cave entrance part) Tinytag air temperature instruments [Gemini data loggers (UK) Ltd., Orion Group of Companies] with precision of 0.1°C were used. For selected periods (see Chapter 5.2.2, Carbon dioxide monitoring) air temperature monitoring was performed at half-hour measurement intervals.

Carbon dioxide volume concentrations in cave air were measured by Vaisala GM70 instruments in ppm. The resolution of CO_2 measurements was 10 ppm with a measuring range between 0 and 5000 ppm and accuracy ±(1.5% of range + 2% of reading). Data were acquired every 30 min or every hour. For graphic analyses, hourly CO_2 data were used with the exception of Predjama Cave (see Chapter 5.2.2, Carbon dioxide monitoring) where half hourly measurements were used.

Temporal measurements of cave air current velocity (see Chapter 5.2.2, Carbon dioxide monitoring) were measured using a Kestrel 4500 device. This instrument continuously measures air temperature, air pressure, humidity, and wind. However, wind measurements are limited, because the device does not detect velocities that are lower than 0.3 m/s. Also, humidity measurements in Slovene caves show values close to 100%, which does not give confident for interpretation.

In Postojna Cave, 16 air temperature monitoring locations are set at hourly measurement intervals (Fig. 4.1). A single external air temperature

Natural and Anthropogenic Impacts on Cave Climates
ISBN 978-0-12-822954-5
https://doi.org/10.1016/B978-0-12-822954-5.00004-4

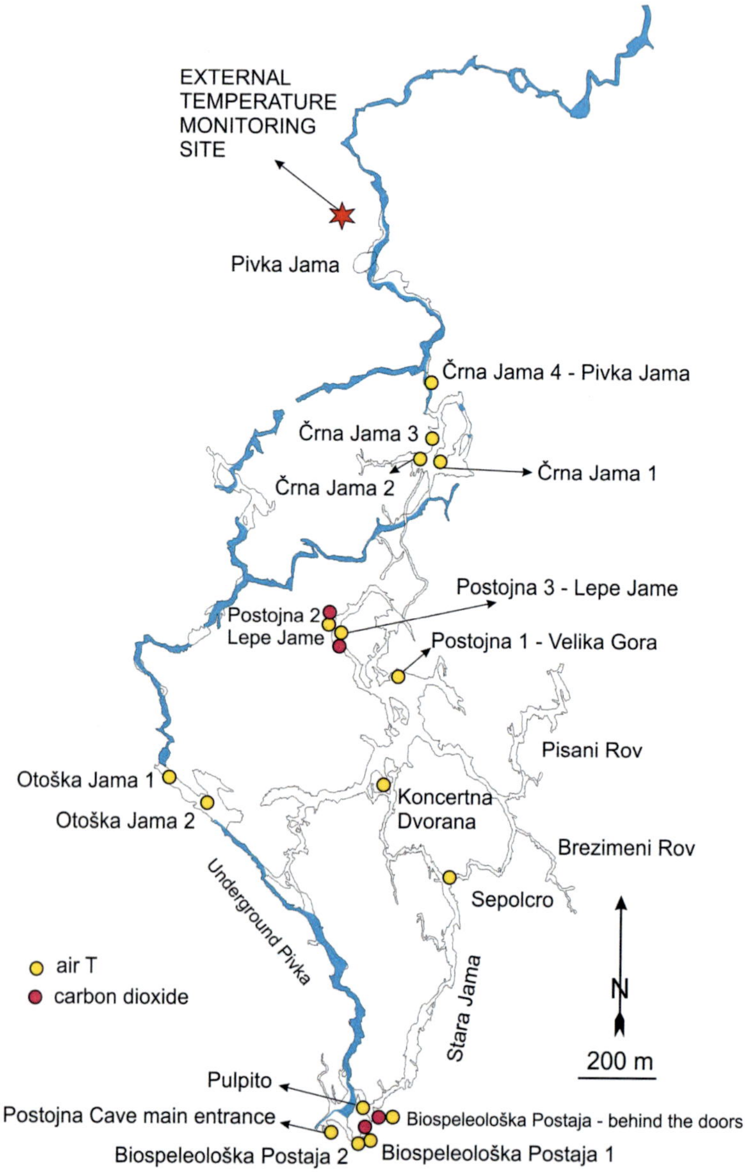

Figure 4.1 Air temperature and carbon dioxide monitoring sites in Postojna Cave and external air temperature site above the cave (red star).

and air pressure monitoring site is placed 559 m above sea level in the forest (Fig. 4.2) close to the entrance to Pivka Jama. The instrument is placed 2.3 m up from the ground. The first air temperature monitoring site in

Figure 4.2 Air temperature data logger in the forest above Pivka Jama. Taking the position of the data logger with a GNSS antenna. *(Photo: S. Šebela.)*

Postojna Cave (Postojna 2 site) started with regular hourly measurements on July 21, 2008 (Šebela and Turk 2011a). The final two temperature data loggers were placed on December 4, 2017, in Otoška Jama (see Chapter 5.1.5, Otoška Jama). The first temperature instruments were placed within passages that were heavily visited by tourists (Lepe Jame, Velika Gora, and Koncertna Dvorana). Later, cave locations were extended to the Biospeleološka Postaja, Črna Jama, Sepolcro, and Pulpito sites. Črna Jama and Otoška Jama monitoring locations were selected as not heavily visited sites. Baro-divers can measure and store hourly measurements data for 1 or 2 years without an electricity supply. Temperature data were downloaded on a portable computer every 3–4 months. On-line real-time temperature data taking is not available with baro-divers. The situation is the same for all other meteorological parameters (carbon dioxide, air circulation, air pressure) used in this study.

Air temperature data sets from continuous measurements in Postojna Cave are not totally complete (especially for Velika Gora and Lepe Jame sites) because problems with instruments and errors in data taking and data reading. On the other hand, the air temperature data set in Predjama Cave is almost complete (see Fig. 5.2.1.5 in Chapter 5.2.1, Air temperature monitoring) because only some measurements are missing at the hourly interval level.

In Predjama Cave, which is not as heavily visited by tourists as Postojna Cave, only two continuous air temperature monitoring locations were organized (see Fig. 5.2.1 in Chapter 5.2, Predjama Cave), while carbon dioxide concentrations were temporarily measured at the Vetrovna Luknja site.

Carbon dioxide measurements are downloaded every 2 months in the case of hourly measurements in Postojna Cave. The Vaisala instrument is connected to electricity. The longest data set of carbon dioxide concentrations in Postojna Cave was taken at Biospeleološka Postaja (see Chapter 5.1.2, Biospeleološka Postaja), where carbon dioxide monitoring started on March 13, 2015.

All meteorological measurements used in graphs for this work are fixed at the UTC +1 time zone (winter time).

The numbers of visitors per year and per month to Postojna (see Chapter 5.1.6 Visitors in Postojna Cave) and Predjama Caves (see Chapter 5.2.3 Visitors in Predjama Cave and castle) were taken from published sources.

Basic statistics (mean, maximum, minimum, range) is calculated for meteorological parameters for different time periods and Pearson correlation coefficients (PCC) are shown for both caves and also for the external monitoring site. Attention is paid in determining the relations between the impact of human presence and the impact of outside climate conditions on cave microclimates.

The PCC or bivariate correlation is a measure of the linear correlation between two variables X and Y. According to Cauchy–Schwarz inequality, it has a value between $+1$ and -1, where 1 is total positive linear correlation, 0 is no linear correlation, and -1 is total negative linear correlation. Pearson's correlation coefficient is the covariance of the two variables divided by the product of their standard deviations.

Cave ground-plans used in this study are taken from the Cave Register of the Republic of Slovenia.

The above sea level measurements (in meters) for different monitoring sites (Lokva ponor at Predjama Cave, paved platform in front of Postojna Cave etc.) were carried out with Leica GS 18T (Fig. 4.2) GNSS antenna (RI-SI-EPOS project).

For all graphs and tables presented in this study, the field meteorological data sets provided by author of this book were used.

CHAPTER 5

Microclimatic monitoring in show caves

In Slovenia, there are 21 active natural karst show caves (Table 5.1), two show caves are closed and an additional three are not actually karst caves but underground mines. Show or commercial caves are caves open for the public that have paths, rails, electric lights, guided tours, and generally have an entrance fee. Show caves are members of the International Show Caves

Table 5.1 Show caves in Slovenia.

Cave	Cave manager	Number of visits (year)
Postojna Cave	Private company	870,000 (2019)
Škocjan Caves	Public institution, regional park	200,000 (2019)
Predjama Cave	Private company	10,000 (2019)
Vilenica	Caving club	9000 (2006)
Divaška Jama	Caving club	n.a.
Dimnice	Caving club	3000 (2007)
Kostanjeviška Jama	Caving club	6000 (2019)
Županova Jama	Tourist association	5000 (2002)
Sveta Jama	Caving club	1000 (2007)
Aragonitna Ravenska Jama	Closed	0
Planinska Jama	Caving club	2000 (2006)
Zelške Jame	Caving club	n.a.
Križna Jama	Association	4000 (2006)
Železna Jama	Caving club	n.a.
Snežna Jama	Caving club	3400 (2006)
Jama Pekel	Tourist association	11,300 (2015)
Rotovnikova Jama	Closed	0
Velika Ledena Jama v Paradani	Tourist association	n.a.
Zadlaška Jama	Caving club	n.a.
Jama pod Babjim Zobom	Caving club	n.a.
Gabrovška Jama	Caving club	n.a.
Huda Luknja	Caving club	n.a.
Francetova Jama	Caving club	n.a.

n.a., not answered.

Natural and Anthropogenic Impacts on Cave Climates
ISBN 978-0-12-822954-5
https://doi.org/10.1016/B978-0-12-822954-5.00005-6

Association (ISCA), which promotes respect and safeguarding of the subterranean environment and its natural wonders, and aims at promoting the economic development of these caves by supplying to all its members (i.e., the show caves) the experiences and solutions gained on the subject of research and management activities carried out by the most advanced caves, which can afford higher expenses (http://www.i-s-c-a.org/objectives-and-purpose).

The most visited show caves including Postojna Cave, Škocjan Caves, and Predjama Cave (Fig. 5.1) have organized systematic microclimatic and biological monitoring. All karst caves in Slovenia are national treasures that belong to the Republic of Slovenia. Most show caves are managed by local caving clubs and some by tourist associations (Table 5.1). Škocjan Caves are protected within a regional park that is managed by the public agency under the Ministry of Environment and Spatial Planning. Škocjan Caves are also the only karst caves in Slovenia that have been listed as a UNESCO's World Heritage Site since 1986. All show caves need a balance between protection of the natural worth and increased tourism. This can only be provided by good knowledge of the natural characteristics of the cave performed by karst research studies, climatic and biological monitoring, and specifying the permissible level of anthropogenic impact.

Vilenica, which is situated in southwest Slovenia (Fig. 5.1), is the oldest show cave in Europe. In 1633 it was opened for guided tours (Kranjc, 2010), and modern visits have been organized since 1962. Vilenica has hosted literary evenings since 1980 that became international in 1986 (Kranjc, 2010). Vilenica (Fig. 5.2) is richly decorated with speleothemes and

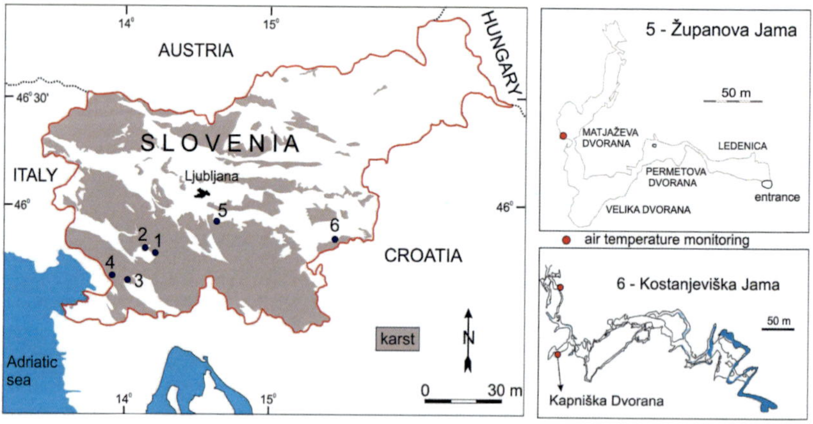

Figure 5.1 Selected show caves in Slovenia and extension of karst area. 1, Postojna Cave; 2, Predjama Cave; 3, Škocjan Caves; 4, Vilenica; 5, Županova Jama; 6, Kostanjeviška Jama.

Figure 5.2 Speleothems in Vilenica Show cave. *(Photo: S. Šebela.)*

hosts numerous shield- or palette-shaped speleothems. Occasional meteorological measurements have been performed in the cave but systematic continuous measurements have not been performed yet.

In some other not massively visited show caves, such as Županova and Kostanjeviška Jama (Fig. 5.1), microclimatic monitoring is more systematic.

The results of air temperature monitoring in Županova Jama showed the relative stability of air temperature in the isolated part of the cave, whereas variable air temperature was observed in the parts close to entrances and the surface (Ravbar and Košutnik, 2014). The distinctive seasonality in cave entrance parts demonstrated that air convection is a driving force for the heat exchange between the cave and the surrounding environment (Ravbar and Košutnik, 2014). External air temperature and heat conducted through the rock walls are also important factors influencing the Županova Jama cave climate, while heat released by ice deposits and water infiltrating through the cave ceiling has a negligible effect (Ravbar and Košutnik, 2014). The entrance chamber (Ledenica) to Županova Jama acts as a cold air trap cave similar to Črna Jama (see Chapter 5.1.4, Črna and Pivka Jama) within the Postojna Cave system. In remote chambers, such as in Matjaževa Dvorana (Fig. 5.3), air temperature monitoring shows stable temperature conditions. The mean annual air temperature in this chamber is 9.5°C and is similar to the mean annual external air temperature. However, during visits to the cave we detected up to 1.5°C higher air temperature peak (Fig. 5.4), which returns back to the previsit values when visitors leave the chamber. To reach the peak, air

Figure 5.3 Air temperature monitoring site (*white arrow*) at Matjaževa Dvorana chamber in Županova Jama. The tourist path is 1.5 m wide. *(Photo: S. Šebela.)*

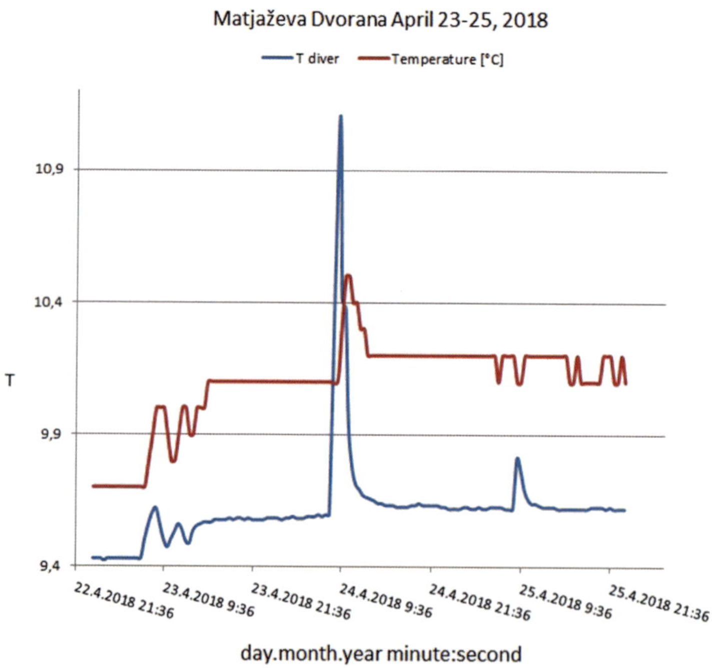

Figure 5.4 Air temperature measured every 30 min with two instruments at Matjaževa Dvorana chamber in Županova Jama April 23–25, 2018, in °C.

temperature increased for 1.5 h. The temperature data logger needed 5 h to recover to the previsit temperature values. Air temperature was measured at the same location every 30 min with two different instruments. A baro-diver was the first and a Radim the second instrument that primarily measures radon concentrations and is placed inside the protective box. The measurements from the baro-diver are more trusted because of the better accuracy and because the baro-diver is placed in the cave air and is not covered with a protective box. The number of visitors was <15 but some of them stayed close to the temperature monitoring site for almost 2 h. Such a situation was discussed also by Ravbar and Košutnik (2014), who wrote that occasional irregular variations in Županova Jama air temperature daily patterns are caused by human impacts.

In Kostanjeviška Jama (Fig. 5.1), two air temperature monitoring sites have been set at hourly intervals since January 16, 2020 (Fig. 5.5). Kostanjeviška Jama 1 site is placed in a remote Kapniška Dvorana chamber ~140 m far from the cave entrance, while Kostanjeviška Jama 2 site is situated ~50 m far from the cave entrance. The instruments are 2–3 m from the tourist path (Fig. 5.6). Kostanjeviška Jama is a refuge for numerous bats (Fig. 5.7), and so the cave is closed for visitors during bat hibernation in winter.

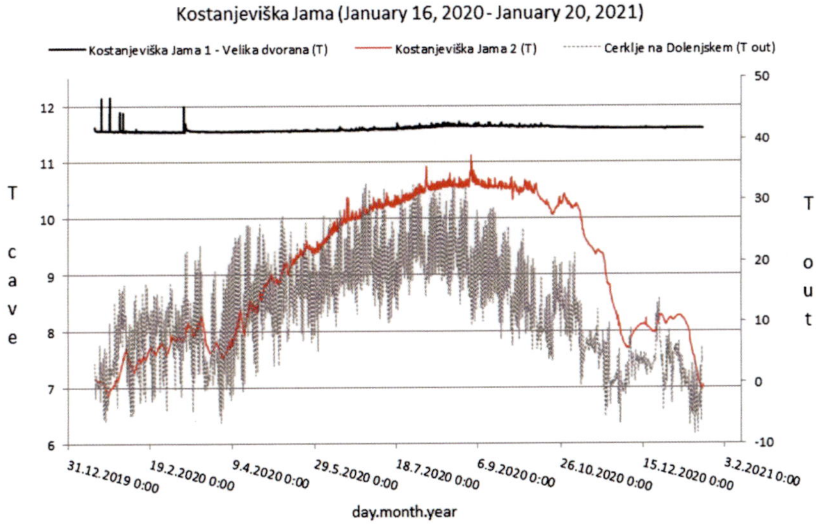

Figure 5.5 Air temperature data set in Kostanjeviška Jama for the period January 16–January 20, 2021, in °C at two monitoring sites. Kostanjeviška Jama 1, Kapniška Dvorana; Kostanjeviška Jama 2, near cave entrance.

Figure 5.6 Kostanjeviška Jama 2 air temperature monitoring site in Kostanjeviška Jama. An air temperature instrument can be found in the top right of the picture. *(Photo: S. Šebela.)*

Figure 5.7 Bats in Kostanjeviška Jama. *(Photo: S. Šebela.)*

The year-round air temperature is stabilized at 11.59°C in Kapniška Dvorana chamber (Fig. 5.5), which is about as high as mean annual external air temperature (11.7°C). Anthropogenic impact from visitors in the period January 2020–January 2021 is not important because the cave received only limited number of people. Close to the cave entrance the air temperature (Kostanjeviška Jama 2 site) is around 9°C (Fig. 5.5). This part of the cave has much a stronger cave climate impact from the winter external climate than Kapniška Dvorana.

We can see that already preliminary meteorological studies in not heavily visited show caves give important information for show cave managers and represent significant added value for sustainable use of caves for tourism.

Microclimatic monitoring of the well-visited Postojna and Predjama show caves will be discussed in detail in the following chapters (see Chapter 5.1, Postojna Cave air temperature and carbon dioxide monitoring and Chapter 5.2, Predjama Cave). Here, we discuss a specific event that took place on November 9, 2017, at Koncertna Dvorana in Postojna Cave. On the occasion of 38 million visitors to Postojna Cave since 1818 (see Chapter 5.1.6 Visitors in Postojna Cave, Fig. 5.1.6.4) a special performance event was organized for visitors at the Koncertna Dvorana chamber (see Figs. 5.1.3.5.1 and 5.1.3.5.2, Chapter 5.1.3.5 Koncertna Dvorana). The event lasted for 1 h (from 10:00 a.m. to 11:00 a.m.) and caused an air temperature increase of ~3°C at the Koncertna Dvorana monitoring site (from 11.2 to 14.3°C). The air temperature returned to the pre-event situation (Fig. 5.8) after 2 h (at 13:00 p.m.). During this event there were about 400–500 tourists in Koncertna Dvorana. Other cave temperature monitoring sites (Fig. 5.8) did not react to this event, because people did

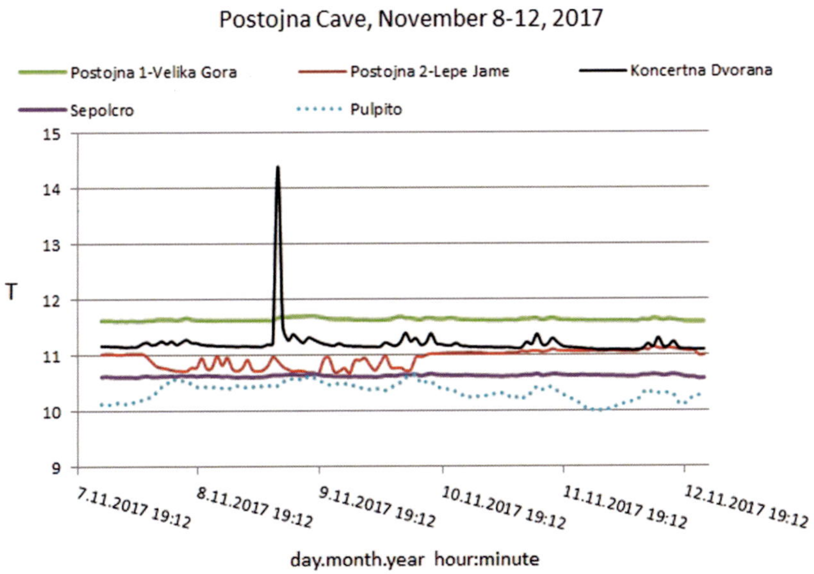

Figure 5.8 Air temperature measurements in Postojna Cave in °C for the period November 8–12, 2017.

not stop for longer periods close to temperature data loggers as was the case in the Koncertna Dvorana chamber. Although the Koncertna Dvorana volume is ~50,000 m^3, a longer presence (at least 1 h) of a larger group with additional lighting effects could temporarily increase the cave air temperature significantly.

When discussing the most visited show caves in Slovenia, Škocjan Caves, as a UNESCO site where more meteorological cave studies have been accomplished in the past, cannot be forgotten (Vercelli, 1931; Kranjc and Opara, 2002; Grgić et al., 2014; Mulec et al., 2017; Debevec et al., 2018). Modern tourism in the Škocjan Caves is considered to have begun on January 1, 1819, when stairs were built that lead to the bottom of the Velika Dolina collapse doline, however, the first documented visits took place as early as 1750 (Kranjc and Opara, 2002).

The most recent microclimatic monitoring started on December 19, 2017, and is still on-going. Škocjan Caves (Fig. 5.9), or Škocjanske Jame in

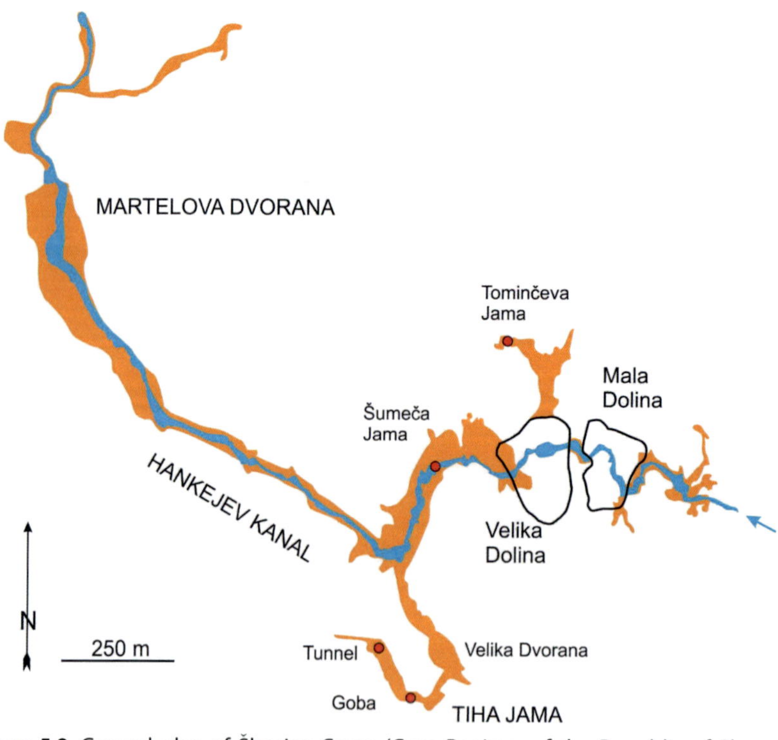

Figure 5.9 Ground plan of Škocjan Caves (Cave Register of the Republic of Slovenia) with air temperature monitoring sites (red dots).

the Slovene language, are 6474 m long and 254 m deep water caves (https://www.jamarska-zveza.si/index.php/strokovne-sluzbe/ss-kataster-jam/ss-kj-prejeto-gradivo/78-kataster-arhiv/47-najdaljse-in-najgloblje-jame-v-sloveniji) on the southeast part of the Kras plateau (Peric and Hribar, 2009). Tourist visits start with an artificial tunnel at the southwest side of the cave and continue through Tiha Jama or Silent Cave (Fig. 5.9), because in this part the underground River Reka cannot be heard. The tourist path leads to the Cerkvenik bridge (Fig. 5.10) to Šumeča Jama or Effervescent Cave where the noise of the river in the underground canyon is very loud (Peric and Hribar, 2009). Hankejev Kanal or Hanke channel is not included in regular tourist visits but it can be visited as an alternative tour. Martelova Dvorana or Martel's chamber in Hankejev Kanal is the second largest underground chamber in Europe with a volume of 2.55 million m^3 (Walters and Zupan Hajna, 2020).

Figure 5.10 Cerkvenik bridge is constructed 48 m above River Reka in the underground canyon of Škocjan Caves. *(Photo: S. Šebela.)*

Four air temperature monitoring sites are situated in Škocjan Caves (Fig. 5.9) with measurements taken at hourly intervals (Fig. 5.11). External air temperature values are taken from archive measurements of the Ministry of the Environment and Spatial Planning, Slovenian Environmental Agency (https://meteo.arso.gov.si/met/sl/archive/).

In a meteorological sense, Škocjan Caves have two distinctive parts. The first is Tiha Jama, which represents a dry passage and the second is Šumeča Jama, which is an underground water canyon. We distinguish summer ($T_{out} > T_{cave}$) and winter ($T_{out} < T_{cave}$) ventilation regimes. The tunnel monitoring site is placed just after the end of the artificial tunnel, which was constructed in 1933, and through which groups of visitors enter the cave. During the high visitor season (summer ventilation, Fig. 5.11) door opening causes intrusion of warm outside air and heating of the cave air at the tunnel monitoring site for up to 17.13°C (Table 5.2), as was the case in 2019. Goba site in Tiha Jama exhibits more stable cave conditions than the tunnel (Table 5.2). Šumeča Jama monitoring site is subjected to the greatest impact of outside cave meteorological conditions (Fig. 5.11). Tominčeva Jama is not included in regular visits and has ~3°C lower air temperature values than the Goba site. The Goba and tunnel sites (Table 5.2) have comparable mean annual air temperatures with the outside situation.

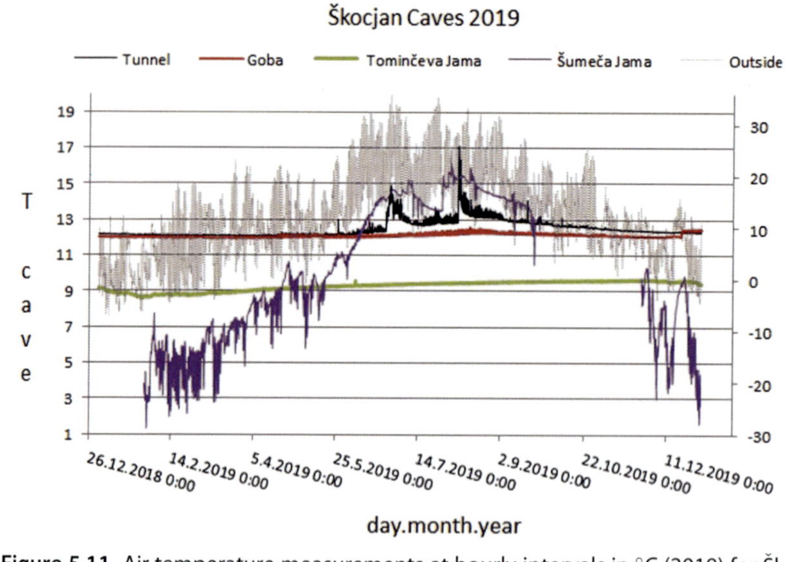

Figure 5.11 Air temperature measurements at hourly intervals in °C (2019) for Škocjan Caves and outside cave monitoring sites.

Table 5.2 Basic statistics for Škocjan Caves air temperature monitoring sites and outside temperature in °C for 2019.

	Tunnel	Goba	Šumeča Jama	Tominčeva Jama	Outside
Mean	12.5	12.11	n.a.	9.28	12.08
Max	17.13	12.57	16.15	9.66	36.1
Min	11.97	11.95	1.31	8.59	−6.9
Range	5.16	0.62	14.84	1.07	43.00

n.a., not available because of the lack of the whole-year data set due to periodical problems with the data logger.

The results of the temperature measurements in the Škocjan Caves (1997–99) have shown that the section of the cave through which the Reka River flows is distinctly dynamic and that visitors do not have much of an impact on the cave meteorology (Kranjc and Opara, 2002). Air temperature measurements in Tiha Jama (Fig. 5.11) showed that door opening between the artificial tunnel and natural cave can greatly impact air temperature at the tunnel monitoring site.

Air temperature monitoring in massively and less heavily visited show caves in Slovenia points to short- and long-term anthropogenic impacts on cave microclimates due to increased visitor numbers and increased presence of people close to the air temperature data loggers. Air temperature monitoring sites in show caves indicate a dependence on outside cave climate conditions. In addition to the studied show caves there are also cave chambers (such as Matevževa Dvorana in Županova Jama, Kapniška Dvorana in Kostanjeviška Jama and northwest part of Tominčeva Jama) that have very stable microclimate conditions with less of a connection with the exterior.

CHAPTER 5.1

Postojna Cave air temperature and carbon dioxide monitoring

Before we present the results of the actual microclimatic in situ observations (see Chapters 5.1.1—5.1.5) we need to introduce past studies on the Postojna Cave microclimate.

The question of what the impact of the presence of visitors on Postojna Cave air temperature has already been asked by Schmidl (1854). From 22:00 p.m. on September 14 to 22:00 p.m. on September 15, 1852, Schmidl (1854) measured the air temperature every hour at the base of the Velika Gora collapse chamber. The mean air temperature was 8.5°C, but when a group of visitors with torches arrived on September 15, 1852, the temperature increased by 0.25°C and remained at 8.75°C until the end of the measurements on September 15, 1852 (Schmidl, 1854).

The work of Crestani and Anelli (1939) entitled *Studies of underground meteorology in the Postojna Cave* written in Italian, composed of 162 pages with numerous figures, graphs, and tables is the most complex monographic work on the Postojna karst meteorology. At more Postojna Cave locations, air temperature, rock temperature, humidity, air pressure, wind speed, wind direction, water temperature, and the level of the underground River Pivka w measured at different periods between 1932 and 1937 (Crestani and Anelli, 1939). Additionally, exterior air temperature in front of the cave and the level of the River Pivka before entering the Postojna Cave were measured also (Crestani and Anelli, 1939).

New speleological connections between Postojna Cave and outside were found, showing important relations between outside cave climate conditions and those in the cave (Crestani and Anelli, 1939). Summer and winter ventilation regimes with transition periods have been determined (Crestani and Anelli, 1939), and cold and warm air-traps within the Postojna Cave system have been described (Crestani and Anelli, 1939).

During meteorological observations in the 1930s, Crestani and Anelli (1939) noted a slight increase in air temperature during traditional feast days. Especially on August 15—16, 1936, and August 14—15, 1937, the increase was at least 0.5°C and it did not return to previsit conditions (over 1000 visitors in 1 day) but lasted the entire day (Crestani and Anelli, 1939).

Natural and Anthropogenic Impacts on Cave Climates
ISBN 978-0-12-822954-5
https://doi.org/10.1016/B978-0-12-822954-5.00006-8

Sporadic measurements of air temperature at the base of Velika Gora in the period from February 1933 to June 1933 showed a mean temperature of 8.6°C (Crestani and Anelli, 1939). Comparing this measurement with September 1852 (Schmidl, 1854) we can see that air temperature did not rise in the period 1852 (8.5°C) to 1933 (8.6°C). It must not be forgotten that in 1866 the main Postojna Cave entrance (see Chapter 5.1.1, Postojna Cave entrance part) was modified, thus permitting greater intrusion of exterior air to the cave than before 1866.

Air temperature measurements including the direction of air circulation were taken in the area of blowholes above the passages of Postojna Cave (Crestani and Anelli, 1939), especially in winter when show melted due to the warm cave air blowing to the surface through a number of openings. Air pressure equilibrated warmer external air sunk to the cave through more than 10 known blowholes during summer (Crestani and Anelli, 1939).

In the period after the Second World War, Professor Gams (1968, 1970, 1974) studied the Postojna Cave meteorology. In 1970 the air temperatures in the main Postojna Cave passage and also in the side passages were described as between 8.2 and 9.2°C (Gams, 1970).

From the characteristics of air circulation it is possible to get information about the characteristics of a karst underground space (Gams, 1970). In the case of a cave system with more openings, air oscillation is impacted by the surface wind in karst areas of Slovenia, especially cold "Bora" wind and external temperature differences (Gams, 1970). On November 16, 1963, when the external temperature was 23.5°C there was 22 m^3/s of air circulation blowing from the cave's main entrance toward the outside (Gams, 1970).

Postojna Cave system is composed of climatically different passages: static, dynamic, and mixed (Gams, 1970). External air enters the cave through more entrances and breathing holes. The underground River Pivka is an important factor in the cave's inner air circulation (Gams, 1970).

In the years 1963–1965, Gams (1970) performed more measurements of air temperature, humidity, and the direction of air circulation in Postojna Cave. Drying of the cave walls was most intensive in winter, but condensation was greatest in the transient period between winter to summer circulation and summer to winter circulation (Gams, 1970). In Postojna Cave evaporation was greatest in winter, while condensation prevailed in summer (Gams, 1970). An exception was passage of the underground River Pivka where condensation was greater than evaporation at all times (Gams, 1970).

Gams (1986) demonstrated different types of cold caves and described that, in the case of Črna Jama in Postojna Cave, cold outside air can invade the cave not only in winter but anytime when the outside air temperature is lower than the cave air temperature. This was evident during low outside temperatures at night, which was exacerbated during cold nights in the warmer part of the year as outside air could enter Črna Jama (Gams, 1986). Gams (2004) described air temperatures in Črna Jama that can drop below zero during very cold winters.

From April 1963 to the end of March 1964, carbon dioxide concentrations in Postojna Cave were measured every 2 weeks at three locations (Gams, 1968). It was found that the highest values (up to 1500 ppm) were detected in the summer and autumn periods, while lower concentrations (300 ppm) were measured in winter and spring (Gams, 1968).

In 1972, three field CO_2 measurement campaigns were carried out by Gams (1974). It was found that the highest CO_2 concentrations (>20,000 ppm) were measured in passages not heavily visited by tourists, such as in Pisani Rov (Gams, 1974). The higher CO_2 concentrations were due to a better connection of the cave air with surface soil CO_2 through the thinner cave ceiling (Gams, 1974).

When cold winter air enters Postojna Cave it is warmed by the cave air. This causes drying of wet cave walls and speleothemes, which can be proven by air temperature changes (Gams, 2002).

Postojna Cave waters (pool, drip water) showed a very constant temperature over a 1-year period, which was very close to the mean outdoor air temperature (Vokal et al., 1999).

In 2009, the International Karstological School "Classical Karst" Conference (Gabrovšek and Mihevc, 2009) was dedicated to cave climate. Therefore, 2009 can be described as the beginning of modern microclimate studies in Postojna Cave because after this year the most intensive microclimate studies were commenced.

The first systematic, automatic, and continuous air temperature and air pressure measurements started in Postojna Cave (Postojna 2 in Lepe Jame, see Fig. 4.1 in Chapter 4, Methods) on July 21, 2008 (Šebela and Turk, 2011a). Monitoring was later extended to the Velika Gora—Postojna 1 (March 18, 2009) and Lepe Jame—Postojna 3 (May 6, 2009) sites (Šebela and Turk, 2011a, b). It was confirmed that air pressure at three monitoring cave sites (Postojna 1, 2, and 3) fluctuated synchronically with the outside air pressure. Temperature data for the Postojna 1 and 3 sites showed a good correlation with outside climate conditions for the period 2009–10, while the seasonal trend at Postojna 2 did not match to outside trend (Šebela and Turk, 2011b). Especially in winter, the Postojna 2 site showed local variability within the rest of the cave system, exhibiting special microclimate conditions that may be due to ventilation of unknown passages at the back (Šebela and Turk, 2011b).

The temperature of the percolating water in Postojna Cave was found to be 9.3–9.8°C (2003–04) and 9.4–9.8°C (2004–05) (Kogovšek, 2010). The measurements showed a seasonal oscillation of percolating water temperatures, with the highest values from October–December and the lowest values from February–April, indicating a 3- to 4-month time lag behind the air temperature of the surface (Kogovšek, 2010).

The relative importance of convection, conduction, and radiation as a function of time was studied for the underground River Pivka channel (Covington et al., 2011). The underground River Pivka undergoes constant temperature fluctuations, with consistent diurnal variations frequently exceeding 1°C (Covington et al., 2011). In the case of karst conduits, essentially all of the heat exchanged must ultimately pass through the rock via conduction (Covington et al., 2011).

In addition to cave air temperature, groundwater temperatures were measured in a rimstone pool in Postojna Cave (Šebela and Turk, 2014a). It was shown that visitors have a greater impact on rimstone pool temperature than on cave air, especially during winter (Šebela and Turk, 2014a).

During summer the amount of heat entering Postojna Cave due to natural ventilation is about twice to 2.8 times higher than the heat transferred to the cave environment by visitors (Šebela and Turk, 2014a).

In preparation for the 200th anniversary of the discovery of new extensive parts of Postojna Cave, a study into cave microclimate and tourism was published (Šebela et al., 2015a, b). Microclimatic and biological studies have been presented to provide guidelines for sustainable use of the cave and to mitigate the side effects of long-term tourism (Šebela et al., 2015a, b).

Carbon dioxide concentrations were included in more studies related to Postojna Cave (Šebela et al., 2013; Gregorič et al., 2013; Mandić et al., 2013; Krajnc et al., 2017; Prelovšek et al., 2018).

In Postojna Cave a 1-day increase in visitor numbers from 200 to 3500 raised the cave air temperature by a maximum of 0.5°C, and a fivefold increase in visitor numbers resulted in a carbon dioxide concentration that was at least twice as high as normal (Šebela et al., 2013).

The CO_2 concentrations of the air in Postojna Cave (400–7900 ppm) were found to be induced by different CO_2 sources, such as human respiration, outgassing of dripwater, water seeping from the vadose zone/epikarst, from the underground Pivka River, and from inflow of outside air with a higher CO_2 concentration (Prelovšek et al., 2018). Some dead-end passages in Postojna Cave with less effective ventilation had cave air CO_2 concentrations of up to 7900 ppm (Prelovšek et al., 2018).

Radon concentrations (Vaupotič et al., 1998, 2001; Vaupotič, 2008; Šebela et al., 2010) and air temperatures were studied in Postojna Cave (Gregorič et al., 2011, 2014; Domínguez-Villar et al., 2015). The outside air temperature was found to have the highest effect over cave ventilation (Gregorič et al., 2014). The radon concentration was in the range from 500 to 9000 Bq/m^3 in 2010, and 500–5500 Bq/m^3 in 2011 and 2012 (Gregorič et al., 2014). A significant change in the radon emissions was observed after high floods in 2010 (Gregorič et al., 2014).

Inside the Postojna Cave, several automatic meteorological measuring stations have been set up, as an adjunct to the surveillance infrastructure; the regular data transfer provided by the DTN (Delay- and Disruption-Tolerant Networking) technology allows the surveillance system to take on the role of an early warning system (EWS) (Gabrovšek et al., 2014).

Postojna Cave has a very complex arrangement of microclimate monitoring locations, because it is a large cave system with different microclimate characteristics. Twenty cave microclimate monitoring locations (Table 5.1.1, see Fig. 4.1 in Chapter 4, Methods) have been analyzed in this study. Almost all the monitoring sites cover hourly air temperature measurements. Additionally, some sites provide periodical carbon dioxide measurements. The outside air temperature measurement in the forest above Pivka Jama was organized by the author. Precipitation data (Slovenian Environmental Agency, https://meteo.arso.gov.si/met/sl/archive/) were used for selected events.

Table 5.1.1 Micro-climate monitoring locations in Postojna Cave and outside.

Monitoring site	Air T	Carbon dioxide	Precipitation
Postojna Cave main entrance 0.20 m	✓		
Postojna Cave main entrance 3.60 m	✓		
Postojna Cave main entrance 6.20 m	✓		
Biospeleološka Postaja 1	✓		
Biospeleološka Postaja 2	✓		
Biospeleološka Postaja near glass plate		✓	
Biospeleološka Postaja—behind the doors	✓	✓	
Pulpito	✓		
Velika Gora	✓		
Postojna 3—Lepe Jame	✓	✓	
Postojna 2—Lepe Jame	✓	✓	
10 m E from Postojna 2—Lepe Jame		✓	
Koncertna Dvorana	✓		
Sepolcro	✓		
Črna Jama 1	✓		
Črna Jama 2	✓		
Črna Jama 3	✓		
Črna Jama 4—Pivka Jama	✓		
Otoška Jama 1	✓		
Otoška Jama 2	✓		
Outside	✓		✓

CHAPTER 5.1.1

Postojna Cave entrance

From October 2012 to January 2014, air temperature monitoring was organized at Postojna Cave main entrance at hourly intervals. We used Tinytag instruments [Gemini data loggers (UK) Ltd., Orion Group of Companies] which have a precision of 0.1°C. Three instruments were placed on forged doors at 0.2, 3.6, and 6.2 m above the floor on the southern edge of the cave wall (Fig. 5.1.1.1).

The intention was to understand year-round fluctuations between outside and cave air at Postojna Cave entrance (Fig. 5.1.1.2). It needs to be stressed that the principal Postojna Cave entrance was naturally filled with

Figure 5.1.1.1 Installation of upper two air temperature instruments on forged metal doors at Postojna Cave main entrance, 3.6 m and 6.2 m above the cave floor. *(Photo: S. Šebela.)*

Natural and Anthropogenic Impacts on Cave Climates
ISBN 978-0-12-822954-5
https://doi.org/10.1016/B978-0-12-822954-5.00007-X

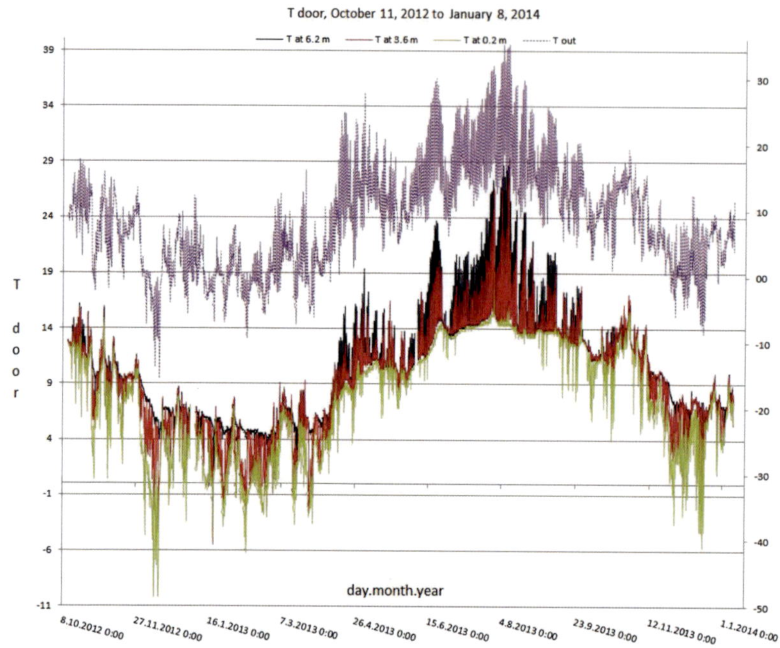

Figure 5.1.1.2 Air temperature hourly measurements in °C at Postojna Cave entrance site from October 11, 2012 to January 8, 2014. Temperature data loggers were placed at 6.2 m, 3.6 m and 0.2 m above the cave floor.

fluvial sediments until the year 1866 (Šebela et al., 2015a, b). With the development of modern tourism after 1818, major modifications to the cave entrance have been performed. In this sense, fluvial sediments were removed in 1866 and the "new" hollow entrance is still today used for visitors.

If we compare outside cave air temperature values with measurements at three elevations (6.2, 3.6, and 0.2 m above the floor) on the forged metal doors for the period October 11, 2012 to January 3, 2013 it is clear that the air temperature at the highest point is between 4 and 13°C in the winter period (Fig. 5.1.1.3). The lowest point, at the cave entrance floor (Fig. 5.1.1.4), has similar temperature values to the outside air (−10 to 13°C). In winter ventilation mode ($T_{\text{out}} < T_{\text{cave}}$) cold outside air sinks into the cave at the bottom of the main entrance. In contrast, at the top of the main cave entrance relatively warm cave air exits the cave during the winter ventilation. This is caused by convective heat transfer or convection, which is the transfer of heat from one place to another by the movement of air masses.

The temperature difference between the highest and lowest placed data loggers was 10.5°C on February 11, 2013 (between 2:00 and 3:00 a.m.).

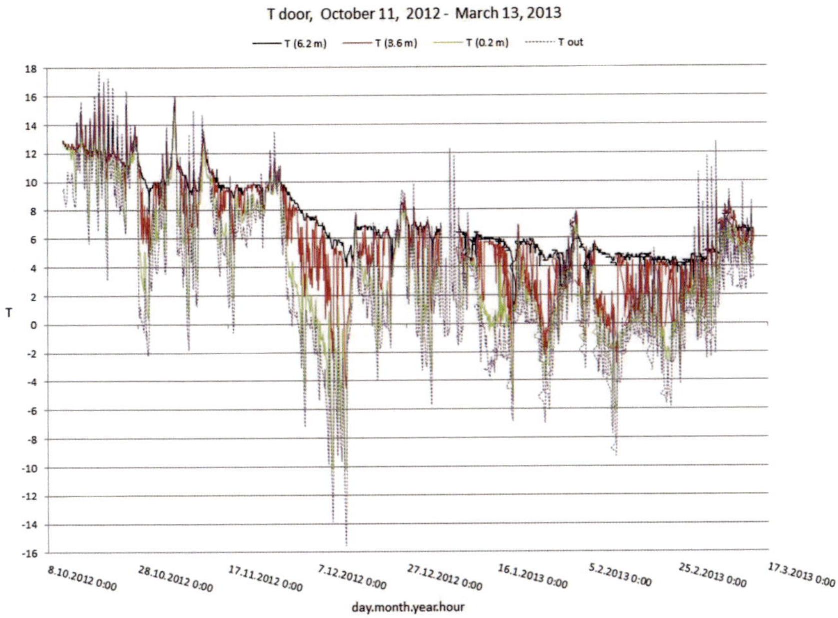

Figure 5.1.1.3 Winter air temperature behaviour °C at Postojna Cave entrance site compared with outside cave air temperature for the period October 11, 2012 to March 13, 2013.

Figure 5.1.1.4 Air temperature data logger placed 0.2 m above the cave floor at Postojna Cave main entrance. *(Photo: S. Šebela.)*

In winter, daily air temperature oscillations are lower at the top of the main entrance than they are at the bottom (Fig. 5.1.1.3).

In the warmer part of the year [summer ventilation ($T_{out} > T_{cave}$)] temperature oscillations are higher at the top and in the middle of the cave entrance than at the bottom of the cave entrance (Fig. 5.1.1.5). In summer, relatively colder cave air exits the cave at the bottom of the main entrance. August measurements of air temperature at the top and middle of the cave's main entrance showed temperatures as high as 27–29°C (Fig. 5.1.1.5), which shows that warm outside air enters the cave (Fig. 5.1.1.6). Part of this warm air returns back to the cave entrance and only part of the warm air proceeds toward internal cave passages through the Veliki Dom chamber.

Basic statistics, including mean annual air temperature values at different heights of the cave entrance, and minimum and maximum air temperature values compared with outside cave data, are presented in Table 5.1.1.1. These data were collected over a year between January 7, 2013 and January 7, 2014. Mean annual air temperature at the middle of the cave entrance doors is similar to outside the cave, which is reasonable due to the fact that we are monitoring the boundary area between the "real" cave and the outside cave environment. The greatest difference between the maximum and minimum air temperature measurements ($\Delta T = 29°C$) during this 1-year period was detected at the top of Postojna Cave main entrance at 6.2 m, while the lowest difference ($\Delta T = 25.5°C$) was at 0.2 m. Airflow fluctuations increase with ΔT.

At Postojna Cave main entrance, summer and winter ventilations are opposites in the sense of air circulation direction. Winter ventilation, due to cold and dense air entering the cave, is well expressed deep inside the cave.

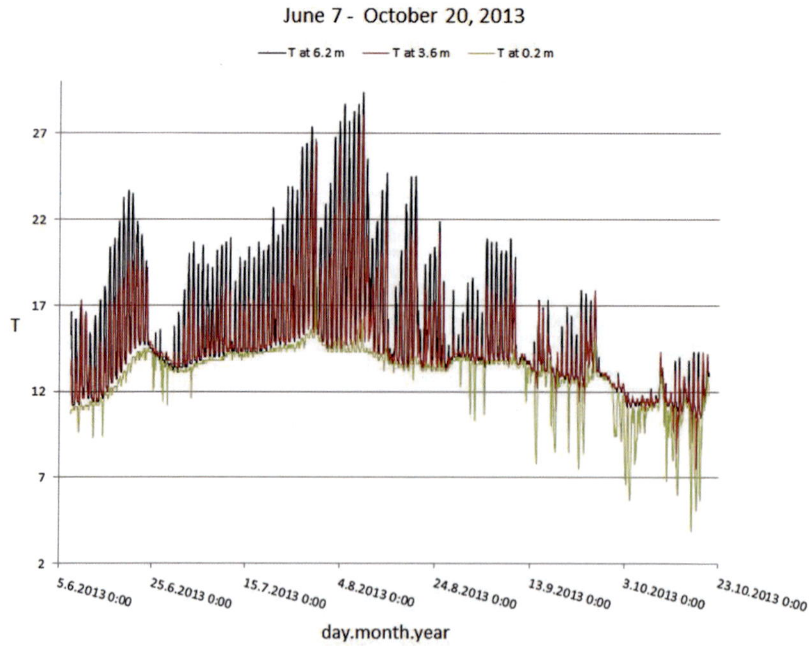

Figure 5.1.1.5 Summer air temperature (in °C) behaviour at Postojna Cave entrance, June 7 to October 20, 2013. Temperature data loggers were placed at 6.2 m, 3.6 m and 0.2 m above cave floor.

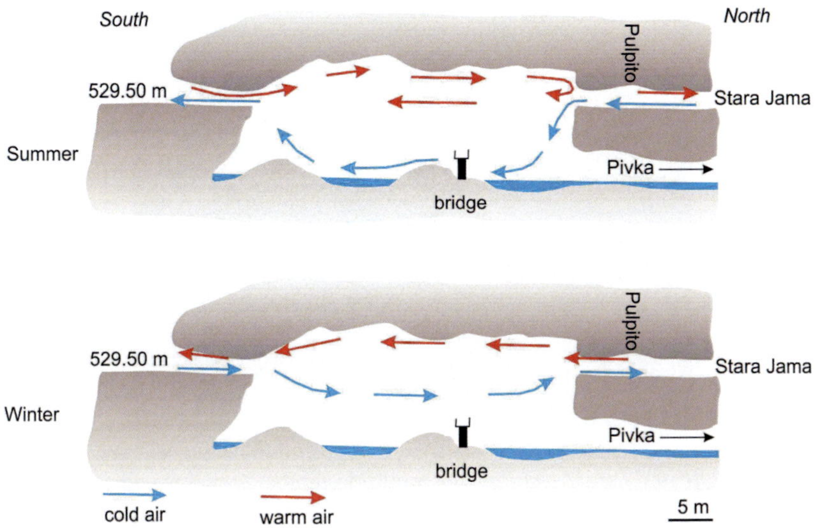

Figure 5.1.1.6 Air circulation of warm and cold air during summer and winter at Postojna Cave main entrance and at Veliki Dom chamber.

Table 5.1.1.1 Basic statistics for air temperature (°C) at Postojna Cave main entrance and outside cave, January 7, 2013 to January 7, 2014.

	Temperature at 6.2 m	Temperature at 3.6 m	Temperature at 0.2 m	Outside temperature
Mean	10.8	9.8	8.1	9.4
Min	0.4	−5.5	−6.3	−9.3
Max	29.4	28.1	19.2	35.3
Range	29.0	33.6	25.5	44.6

As was described in February 2012 (Šebela et al., 2015a, b), the cold northern wind, called "Burja," lowered the cave air at Lepe Jame passage (Postojna 3 monitoring site) by 0.3°C. The average wind speed outside the cave was 3.8 m/s, with a maximum speed of 18 m/s. This was one of the most significant drops in air temperature at Lepe Jame between 2009 and 2013 (Šebela et al., 2015a, b). Even if summer ventilation is not as strong as winter it should not be underestimated (Prelovšek et al., 2018).

Very cold and dense outside air that enters the cave in winter (air temperature below 0°C) at the bottom of the entrance passage causes freezing in some parts of the cave entrance. Breakdown of the cave wall in Veliki Dom (and also in Otoška Jama) was reported as a consequence of ice melting (Crestani and Anelli, 1939). In recent years with milder winters, we do not register such events any more, however during chilly days in winter we witness ice stalagmites and ice ground layers in Postojna Cave entrance (Fig. 5.1.1.7).

Figure 5.1.1.7 Ice stalactites and stalagmites at Postojna Cave entrance part on February 27, 2018 *(Photo: S. Šebela)*, when outside temperature lowered to −15.9°C.*)*

During maintenance work in Postojna Cave (as was the case in February 2019), forged metal doors are sometimes covered by protective foil to protect workers from the powerful winter ventilation in the entrance (Fig. 5.1.1.8). In the case of construction work inside the cave, dust can travel deep into the cave due to winter ventilation. To limit dust transportation, protective foil is placed within the ventilation path in the cave. Because air temperature monitoring at the cave's main entrance was organized only from October 2012 to January 2014, we can only suspect that the closed doors in February 2019 probably influenced air temperature at the bottom of the doors, which did not decrease as it would normally.

Figure 5.1.1.8 Main entrance to Postojna Cave was closed for more days to prevent cold air entering the cave in February 2019. *(Photo: S. Šebela.)*

The most obvious effect of door closing was that we did not have deposition of ice stalactites and stalagmites as was the case in February 2018. However, it is not easy to compare February 2018 conditions with February 2019 (Fig. 5.1.1.9) due to the lack of measurements at the cave's main entrance. Considering another monitoring site called Pulpito (Fig. 5.1.1.6), we can see that this site shows lower air temperature values in 2018 than in 2019, but this is due to the fact that outside conditions in February 2018 were much colder than in February 2019. Even if in February 2019 the cave entrance doors were not closed, we would probably not have had deposition of ice inside the cave because the outside conditions were not as cold as in February 2018.

For cave management and interpretation of cave meteorology, it is very important to have precise information of the time period when entrance doors are closed or open. In the case of very cold winters, closed entrance doors can prevent the growth of ice stalactites and stalagmites in the Postojna Cave's entrance area to some extent. Closed doors cannot completely prevent cold outside air from entering the cave, because there are other lower entrances and outside air can use opened fissures and openings to enter the cave also. However, closed doors can at least reduce ice deposition in cave entrance.

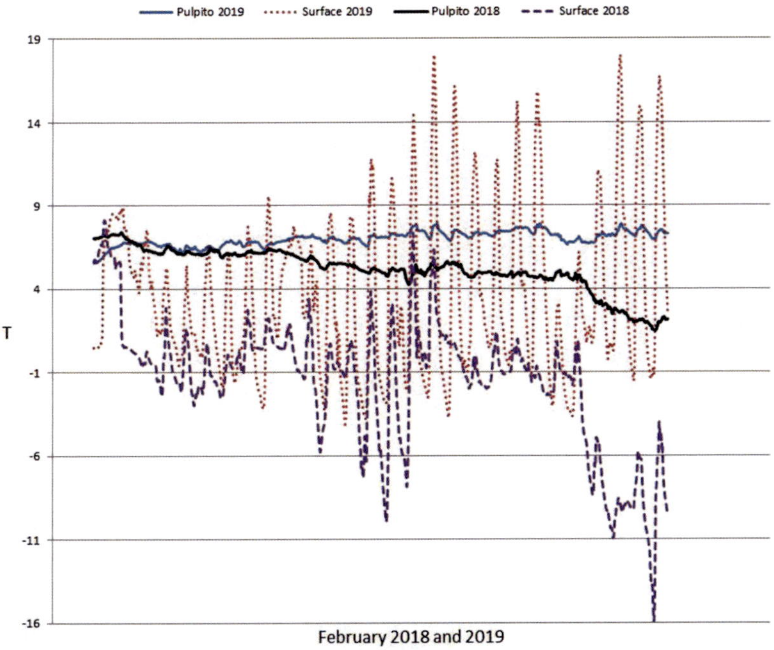

Figure 5.1.1.9 Comparison between air temperature in °C at Pulpito site and outside cave for February 2018 and February 2019.

Veliki Dom, as the first large Postojna Cave entrance room, is characterized also by black coatings covering the cave wall, floor, and flowstone (Zupančič et al., 2011). Such black coatings are from the Second World war when a group of patriots organized an explosion (April 23 to 24, 1944) of aviation fuel that was stored in the entrance areas of the Postojna Cave system. It was reported (Fatur, 1975) that a loud explosion and large cloud of black smoke was seen exiting the principal entrance to the cave; it was still burning 7 days (Fatur, 1975). This information is interesting for cave meteorology because the event happened during summer ventilation, when the outside cave temperature must have been higher than the cave temperature, and black smoke was coming out of the cave.

However, this black coloring had already been described in 1939–40 (Politi) in entrance areas of Postojna Cave. The origin of the coloring was attributed to smoke, dust originating from outside the cave, and also to illumination used during the 19th century (Politi, 1939–40). This means that the black coloring of the cave entrance was present before the explosion in 1944, however the explosion may have add intensity and more pollution to the already existing black coloring.

Additionally, we must not exclude historic and prehistoric fires in the area around Postojna Cave (Šebela et al., 2017) as possible causes of the black coloring in the cave's entrance areas. The oldest part of Postojna town was destroyed by numerous fires in the past. In the last great fire, in 1802, almost all of the Old Quarter has burnt down (Habe, 1986). During winter ventilation, smoke, due to fires, can enter the cave and cause black coloring, as described for Črna Jama (Šebela et al., 2017), and cannot be excluded as one of the causes of the dark coloring of Velik Dom chamber also.

CHAPTER 5.1.2

Biospeleološka Postaja

Contents

Biospeleološka Postaja (Biospeleological Station) is a small part of Postojna Cave (Fig. 5.1.2.1), which is offered for visitors separately from the other

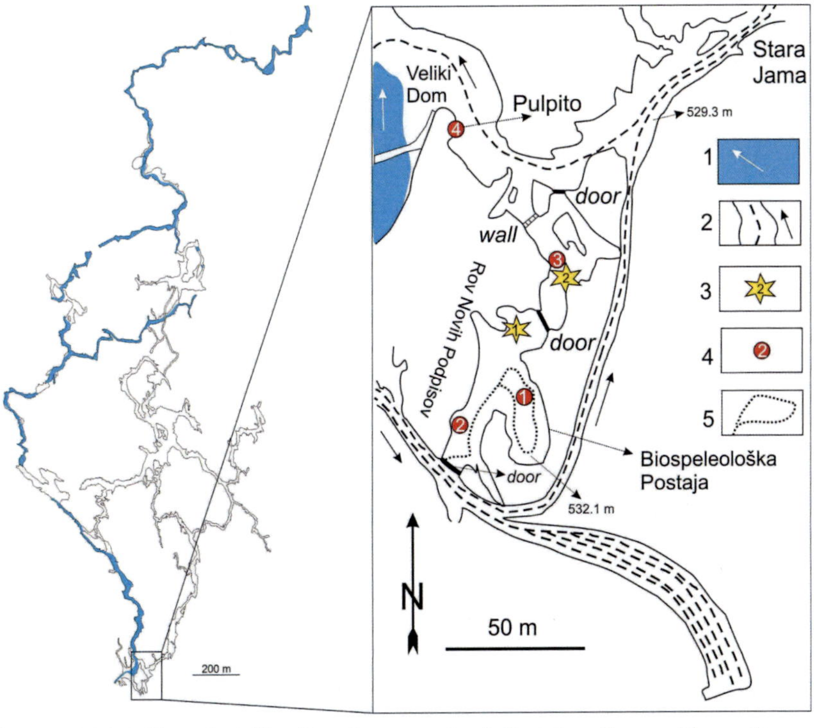

Figure 5.1.2.1 Biospeleološka Postaja as part of Postojna Cave with position of air temperature and CO_2 instruments. 1- underground River Pivka with flow direction, 2- cave raliway with direction of the visits, 3- CO_2 instrument (1=CO_2, 2=CO_2 behind the doors), 4- air temperature instrument (1=Biospeleološka Postaja 1, 2=Biospeleološka Postaja 2, 3=behind the door, 4=Pulpito), 5- walking path.

Natural and Anthropogenic Impacts on Cave Climates
ISBN 978-0-12-822954-5
https://doi.org/10.1016/B978-0-12-822954-5.00008-1

Figure 5.1.2.2 Old signatures on the wall of Biospeleološka Postaja. *(Photo: S. Šebela.)*

most visited parts of the cave. Visitors spend about 30–45 min on a walking tour and learn about cave biology. This part of Postojna Cave is also called Rov Novih Podpisov (Passage of New Inscriptions, Fig. 5.1.2.2). The passage was discovered in 1818 when local cave guide and lamplighter Luka Čeč traversed underground River Pivka in Veliki Dom and found extensive new parts of the cave system. Biospeleološka Postaja (or Vivarij) was opened in 1931 in Rov Novih Podpisov (Čuk, 2008). There are numerous aquariums for terrestrial and water cave fauna. Beside didactic information for visitors (Fig. 5.1.2.3), it is also a location for scientific biological research in an underground biospeleological station.

The characteristics of air temperature and CO_2 concentrations in Biospeleološka Postaja have been monitored in hourly intervals since March 13, 2015. Climatic monitoring started because it was anticipated that a small

Figure 5.1.2.3 Didactic displays in Biospeleološka Postaja, location of Biospeleološka 1 monitoring site. *(Photo: S. Šebela.)*

cave passage, being regularly visited, might show altered air temperature values compared with other cave passages. This cave passage was modified in the 1930s, which changed the natural air circulation. There is an artificial wall (Fig. 5.1.2.1) in the northern part of the passage, which blocks air exchange with the inner cave areas. The most northern part of the passage is closed by a metallic net door (Fig. 5.1.2.4), which allows air exchange with Stara Jama and Veliki Dom (Fig. 5.1.2.1).

In the middle of the passage there is another obstacle—metal doors (Fig. 5.1.2.5). When they are closed, they largely prevent air ventilation with the other passages of the cave system. These doors were put in place in

Figure 5.1.2.4 Northern metallic net door, view from Stara Jama to Rov Novih Podpisov. *(Photo: S. Šebela.)*

Figure 5.1.2.5 Closed metal doors in the middle of Rov Starih Podpisov passage, view towards north. *(Photo: S. Šebela.)*

1931, probably to preserve Biospeleološka Postaja from unattended visits to the underground biological laboratory.

The third doors (the most southern ones) were fitted in 2002, when fluvial sediments that blocked the entrance were removed during construction of the new tourist path (Bavdek, 2003; Kranjc et al., 2007). Archeological research into the cave sediments from this passage proved that cultural layers belonged to Moustérien culture (Dirjec and Turk, 1987). With the installation of sliding doors (Fig. 5.1.2.6), the new entrance or exit to Biospeleološka Postaja from the south was opened for visitors as an additional tourist attraction within the Postojna Cave system.

In the first year of systematic microclimatic monitoring in Biospeleološka Postaja, high CO_2 concentrations were detected in summer 2015 (Fig. 5.1.2.7). On August 17, 2015, the highest concentration of 4100 ppm was measured. In this period, metal doors (Fig. 5.1.2.5) were closed year-round. Because the middle of August is the period with the

Figure 5.1.2.6 Sliding doors at the southern entrance of Biospeleološka Postaja (Vivarij). *(Photo: S. Šebela.)*

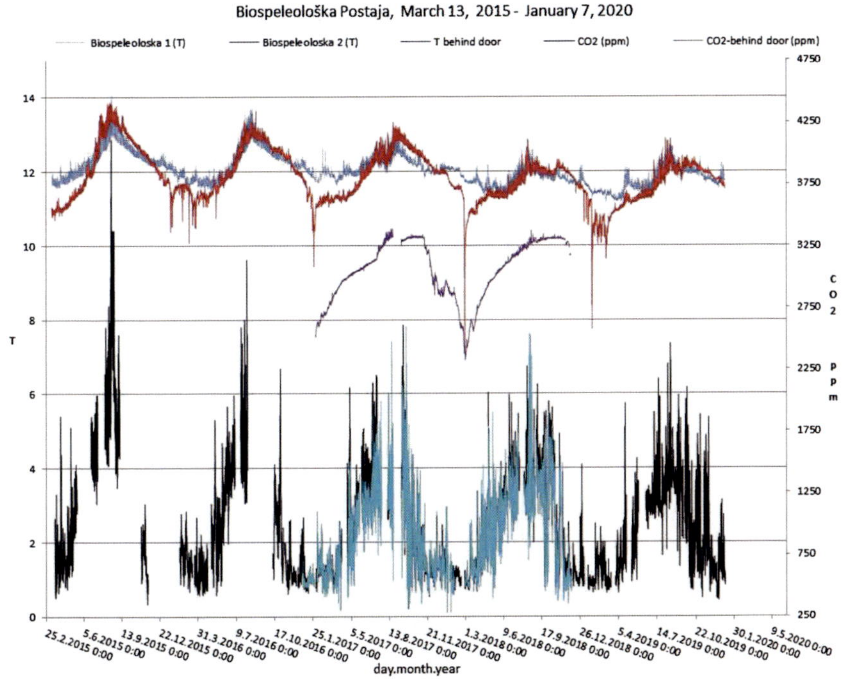

Figure 5.1.2.7 Air temperature (°C) and CO_2 concentractions in Biospeleološka Postaja, March 13, 2015 to January 7, 2020.

highest visitor numbers in the cave, it was obvious that this CO_2 peak was connected with the breathing of visitors in a closed passage with very poor exchange of air with other cave parts. At the same time, the air temperature at Biospeleološka 1 site (Fig. 5.1.2.1) reached 14°C, which set us thinking that we needed to discover a way to lower both parameters.

Atmospheric CO_2 at the Mauna Loa Observatory (https://www.esrl.noaa.gov/gmd/ccgg/trends/) was 406 ppm in October 2018 and 408.53 ppm in October 2019, which was similar to 2015. In caves, there are different sources of CO_2. For Postojna Cave, the CO_2 sources include human respiration, outgassing of dripwater and seeping from the vadose zone or epikarst, CO_2 dilution, and Pivka River (Gregorič et al., 2014; Prelovšek et al., 2018). It can obtain similar concentrations to outdoors mostly in winter, what is also the case with Biospeleološka Postaja. Higher summer concentrations in caves are mostly connected with higher concentrations in the soil above the cave. And even if we think of the summer natural CO_2 background being about 2000–2200 ppm, 4100 ppm in 2015 was excessive. The first proposal for cave management was to open metal doors in summer and the next step was to start additional monitoring at the northern side of the doors. In 2017 and 2018, we operated three air temperature monitoring sites and two CO_2 sites at hourly intervals. We compared microclimatic conditions in a strongly burdened cave passage (south of the metal doors) with a less burdened one (north of the metal doors).

Air temperature monitoring

Fig. 5.1.2.7 demonstrates the air temperature and CO_2 values from March 13, 2015, to January 7, 2020, for Biospeleološka Postaja. As already mentioned, in 2015, the summer temperature and CO_2 values were the highest. In 2016, the summer values dropped because the metal doors (Fig. 5.1.2.5) were generally opened. In 2017, metal doors were still occasionally closed, resulting in the CO_2 instrument behind the doors (number 2 in Fig. 5.1.2.1) showing lower concentrations compared to the instrument south of the metal doors in June 2017. This was the reason why, in 2017, we proposed to the cave management to leave the metal doors opened year round. In fact, the graph for 2017 shows very comparable CO_2 concentrations at both locations and additional lowering of air temperature. In 2018, we stopped measuring the cave air parameters north of the metal doors, as the regulation was successful.

Looking at Fig. 5.1.2.8, it is clear that Biospeleološka 1 monitoring site, situated in the inner part of the chamber (Fig. 5.1.2.3), has lower air temperature fluctuations over a 1-year period (Table 5.1.2.1) than

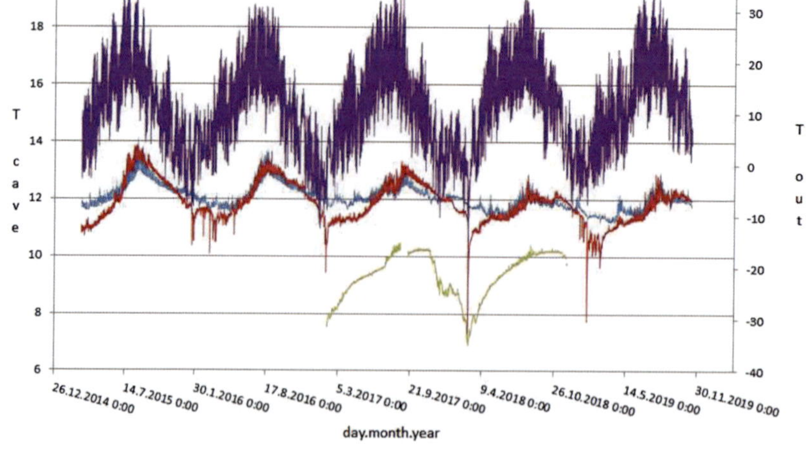

Figure 5.1.2.8 Air temperature (°C) in Biospeleološka Postaja and outside cave for the period March 13, 2015 to November 14, 2019.

Table 5.1.2.1 Basic statistics for air temperature (°C) in Biospeleološka Postaja and outside cave for 2016 to 2019.

	2016—Biospel. 1	2016—Biospel. 2	Outside 2016	2017—Biospel. 1	2017—Biospel. 2	Outside 2017	2018—Biospel. 1	2018—Biospel. 2	Outside 2018	2019—Biospel. 1	2019—Biospel. 2	Outside 2019
Mean	12.19	12.04	9.97	12.11	11.94	10	11.79	11.6	10.55	11.7	11.47	10.58
Max	13.73	13.65	31.31	13.07	13.33	37.07	12.84	12.86	33	12.92	12.89	34.92
Min	11.38	10.08	−9.69	11.64	9.43	−12.31	11.29	7.18	−15.94	11.15	7.73	−7.51
Range	2.35	3.57	41	1.43	3.9	49.38	1.55	5.68	48.94	1.77	5.16	42.43

Biospeleološka 2 site, which is closer to the cave entrance. The mean annual air temperature at Biospeleološka 1 is consistently higher than at Biospeleološka 2 (Table 5.1.2.1). During cold winter days we can see an expressive air temperature drop at Biospeleološka 2 (Fig. 5.1.2.8), which is due to the drop in the outside air temperature (up to −16°C) and break-through of cold and dense outside air into Biospeleološka Postaja, as was the case on February 28, 2018. On the same day, the Biospeleološka 1 site did not respond to the cold outside event and remained with more stable conditions, because it was out of the cold ventilation path. Regarding the third air temperature monitoring site, behind metal doors, the temperature graph (Fig. 5.1.2.8) shows that from July to the beginning of November 2018 the temperature was just a little bit above 10°C. In winter ventilation, when T_{out} was less than T_{cave}, cave air was cooled by the outside cold air entering the cave through Veliki Dom and reaching Biospeleološka Postaja through opened metal doors (Figs. 5.1.2.5 and 5.1.2.9). On the other hand, we cannot exclude outside cold air entering Biospeleološka Postaja at its southern entrance, what is supported by the fact that in the period of February 27 to March 1, 2018, the air temperature behind the metal doors did not drop as much as at Biospeleološka 2 (Fig. 5.1.2.9). We can see that if metal doors are opened during very cold winter days, as was the case in February 28, 2018, and January 27, 2019, this can lower the air temperature at the Biospeleološka 2 site by more than 2°C for about 24 h.

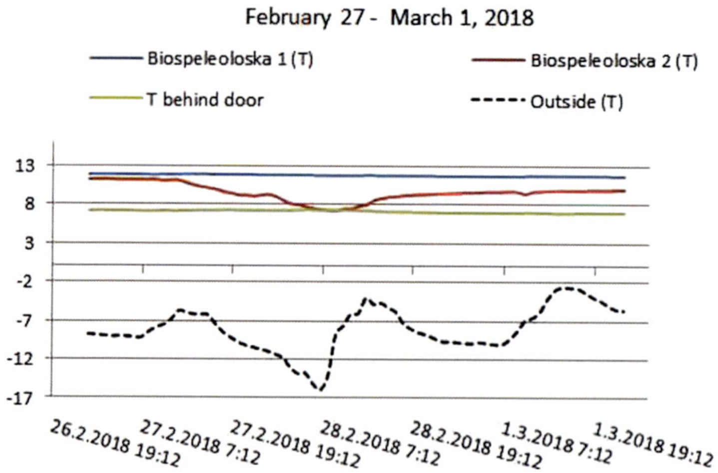

Figure 5.1.2.9 Impact of cold outside air (°C) on air temperature in Biospeleološka Postaja, February 27 to March 1, 2018.

The annual air temperature range at the Biospeleološka 1 site was 2.35°C in 2016, 1.43°C in 2017, 1.55°C in 2018, and 1.77°C in 2019 (Table 5.1.2.1). The reduced temperature range from 2016 to 2019 is due to forced ventilation by the metal doors being opened. At the Biospeleološka 2 site, the annual air temperature range was 3.57°C in 2016, 3.9°C in 2017, 5.68°C in 2018, and 5.16°C in 2019. At this site, the temperature range increased from 2016 to 2019, what is opposite to what happened at Biospeleološka 1. The Biospeleološka 2 site is situated closer to the cave entrance and is more influenced by outside cave air temperature, where annual air temperature range in the period 2016 to 2019 increased, as was the case also for the Biospeleološka 2 site. We cannot exclude a greater cold air influx (in winter), due to metal doors being opened, as an additional cause of the annual air temperature range increase at the Biospeleološka 2 site (Fig. 5.1.2.10).

The mean annual air temperature in Biospeleološka Postaja and outside the cave for the years 2016–2019 (Table 5.1.2.1, Fig. 5.1.2.11) shows a decrease of −0.49°C at Biospeleološka 1, a decrease of −0.57°C at Biospeleološka 2, and an increase of +0.61°C for the outside air temperature. Decreased values at cave sites are due to metal doors being open, as the annual number of cave visitors did not decrease between 2016 and 2019. In 2018, about 130,000 tourists visited Biospeleološka Postaja (https://krog.sta.si/2588193/leto-2018-najboljse-v-zgodovini-turisticnega-obiska-postojnske-jame).

Comparing the mean air temperature for the period 2016 to 2019 between Biospeleološka Postaja and outside (Table 5.1.2.2), it was calculated that the cave air temperature was +1.67°C higher that outside the cave at Biospeleološka 1 and +1.49°C higher at Biospeleološka 2. Such a difference is found between the higher values in the Postojna Cave system and especially between higher values in the tourist-visited parts of the cave. An exception is the occasionally tourist-visited Otoška Jama (monitoring site near the underground River Pivka), which in 2018 showed about a +2°C higher mean annual air temperature compared with the outside cave conditions. At Otoška Jama, we must exclude the impact of visitors on the cave air temperature because the biggest cause of heat transmitted into the cave was by underground River Pivka.

Carbon dioxide monitoring

As already mentioned, in parallel with the air temperature, CO_2 concentrations are studied in Biospeleolološka Postaja at hourly intervals also. At the first CO_2

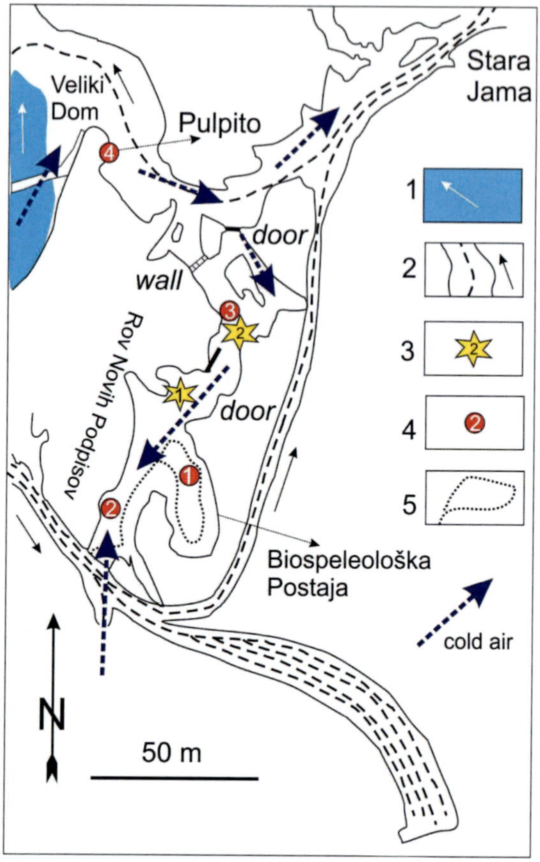

Figure 5.1.2.10 Winter ventilation in Biospeleološka Postaja, February 28, 2018.

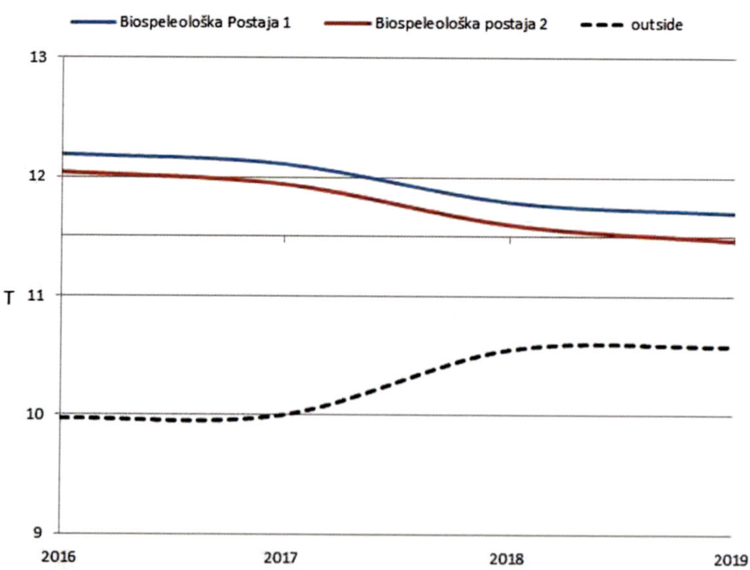

Figure 5.1.2.11 Mean annual air temperatue in Biospeleološka Postaja and outside the cave for 2016 to 2019 (°C).

Table 5.1.2.2 Mean air temperature for 2016 to 2019 in Biospeleološka Postaja and outside cave; the difference (ΔT) between cave air and outside cave air (°C).

	Mean 2016–2019	ΔT (T cave – T out)
Biospeleološka 1	11.94	+1.67
Biospeleološka 2	11.76	+1.49
Outside	10.27	–

monitoring site (Fig. 5.1.2.7), which is situated south of the metal doors (Fig. 5.1.2.1), monitoring has been carried out since 2015. At this site, during periods of high precipitation, percolating water can fill the rimstone pools (Fig. 5.1.2.12). In summer (August 2019), CO_2 concentrations can reach 2200 ppm, while in winter they are around 500 ppm (Fig. 5.1.2.13). Summer concentrations, which are already high due to the transport of CO_2 from the surface soil (active vegetation) to the cave (Prelovšek et al., 2018), are additionally impacted by breathing from higher visitor numbers. An additional source is degassing of CO_2 from percolating water during precipitation, as was detected in November 2019 (Fig. 5.1.2.14). Especially, the CO_2 peak on November 13–14, 2019, is connected with increased outside cave precipitation.

Summer concentrations at the second CO_2 monitoring site (monitoring started on December 28, 2016, to December 3, 2018) behind the doors (with metal doors being opened) can reach 2500 ppm (August 17–23, 2018), which is 1000 ppm higher than the natural background level. This shows that high CO_2 concentrations measured away from the parts of the cave being visited by tourists cannot always be connected to human breathing, but are probably attributed to natural causes within the cave system, such as degassing of CO_2 from percolating water.

Figure 5.1.2.12 Monitoring site for carbon dioxide measurements (2015 to 2020) in Biospeleolološka Postaja in the period of high water (November 14, 2019), view towards south. *(Photo: S. Šebela.)*

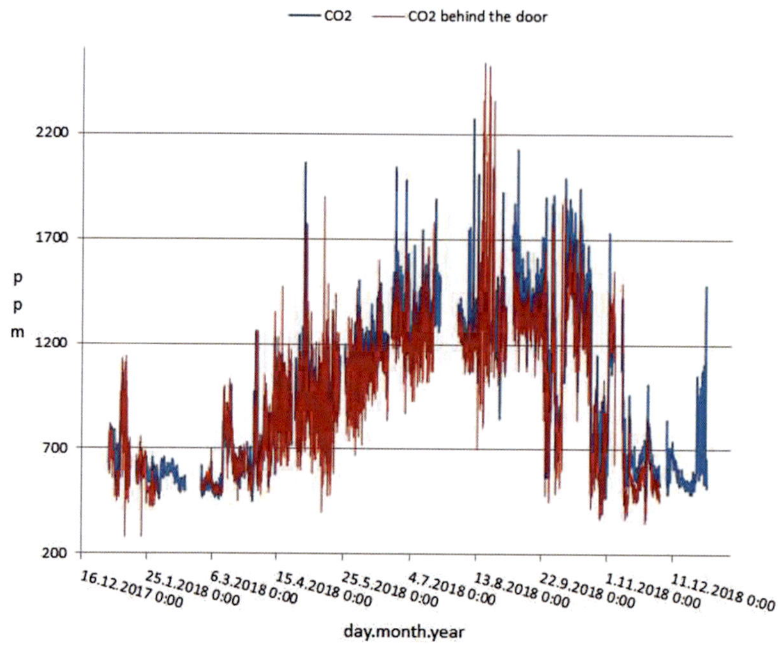

Figure 5.1.2.13 Carbon dioxide concetrations (ppm) at hourly intervals (ppm) at two locations in Biospeleološka Postaja in 2018.

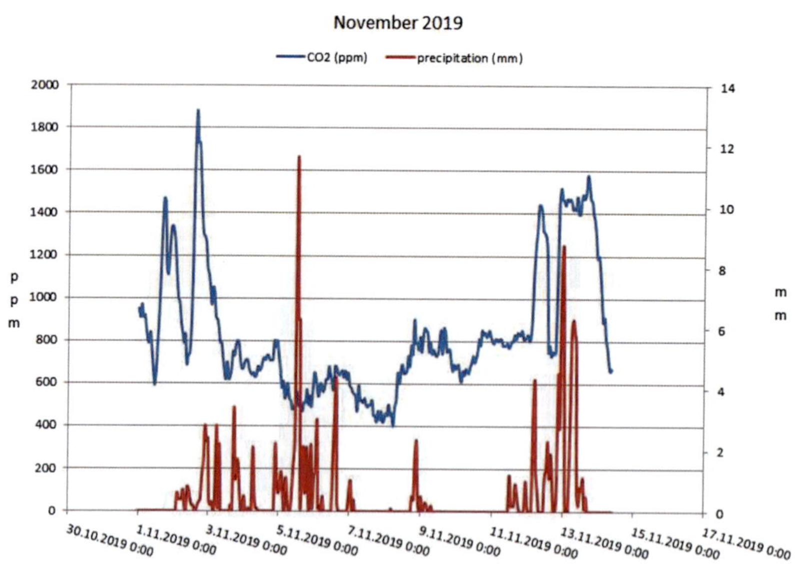

Figure 5.1.2.14 Precipitation in mm per hour (Postojna) and CO_2 concentrations (Biospeleološka Postaja) in November 2019.

Carbon dioxide: noon and midnight concentrations

Tourist visits to Biospeleološka Postaja are, as in other cave parts, limited to daytime hours. When visitors are present in Biospeleološka Postaja, lights are turned on. If there is nobody in the cave the lights are turned off. Comparing day against night cave meteorology conditions can help to see differences or similarities between periods with no visits (night) and period with visits (day).

Comparing midnight (Fig. 5.1.2.15) and noon (Fig. 5.1.2.16) carbon dioxide concentrations at two monitoring sites, we notice that midday CO_2 values reached 2200 ppm in August (2018), while midnight values did not exceed 1700 ppm in the same month. This difference can be attributed to the presence of visitors (breathing).

Daily oscillations of air temperature and CO_2 concentrations (Fig. 5.1.2.17) in the summer period (July 19 to August 7, 2016), when we have the highest number of daily visitors, show the lowest CO_2 concentrations in the early morning (1300 ppm) and the highest values (3100 ppm) in the afternoon. Also, air temperature values are lowest in the morning and highest in the afternoon (around 15:00 p.m.).

The air temperature starts to increase after 7:00 a.m. (Fig. 5.1.2.18) and reaches the highest values at 15:00 p.m., while CO_2 concentrations start to increase after 10:00 a.m. (or after 8:00 a.m.) and reach the highest values at 17:00 p.m. On August 7, 2016, there was a time-lag between the highest daily values of air temperature and CO_2 of 2 h. Generally, this time-lag would be 1 h.

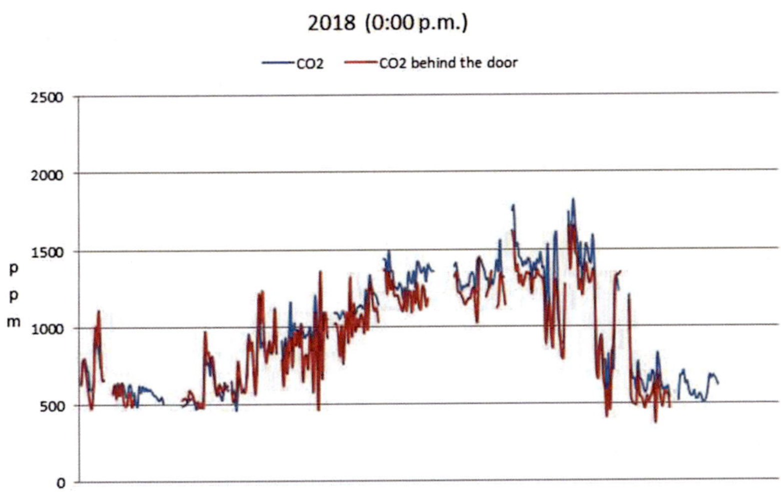

Figure 5.1.2.15 Carbon dioxide midnight (0:00 a.m.) concetrations (ppm) at two locations in Biospeleološka Postaja in 2018.

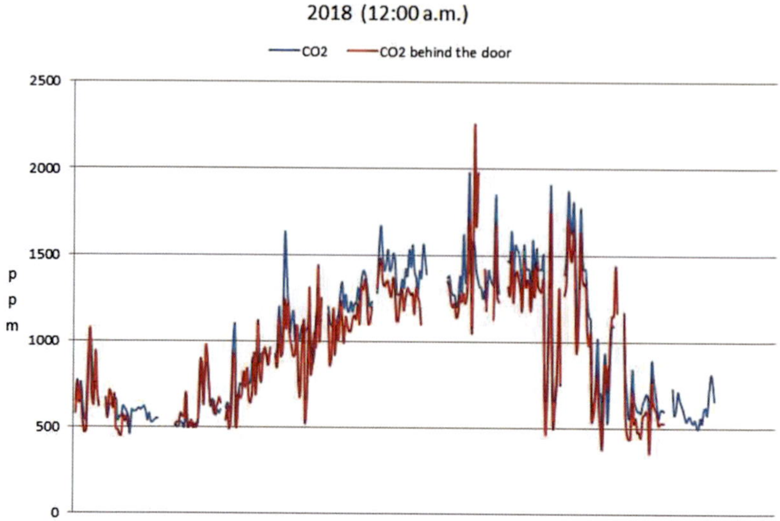

Figure 5.1.2.16 Carbon dioxide noon (12:00 a.m.) concetrations (ppm) at two locations in Biospeleološka Postaja in 2018.

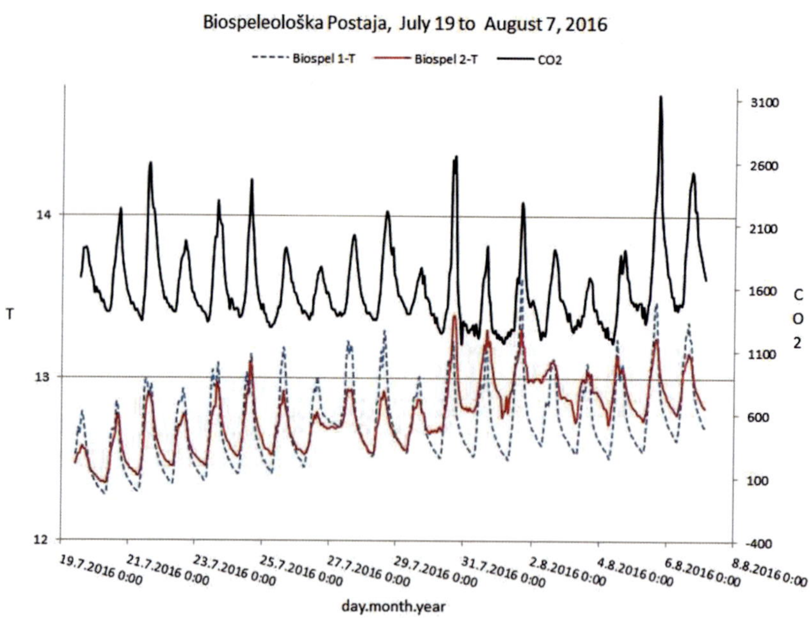

Figure 5.1.2.17 Daily CO_2 concentrations (ppm) and air temperature (°C) oscillations in Biospeleološka Postaja (July 19 to August 7, 2016).

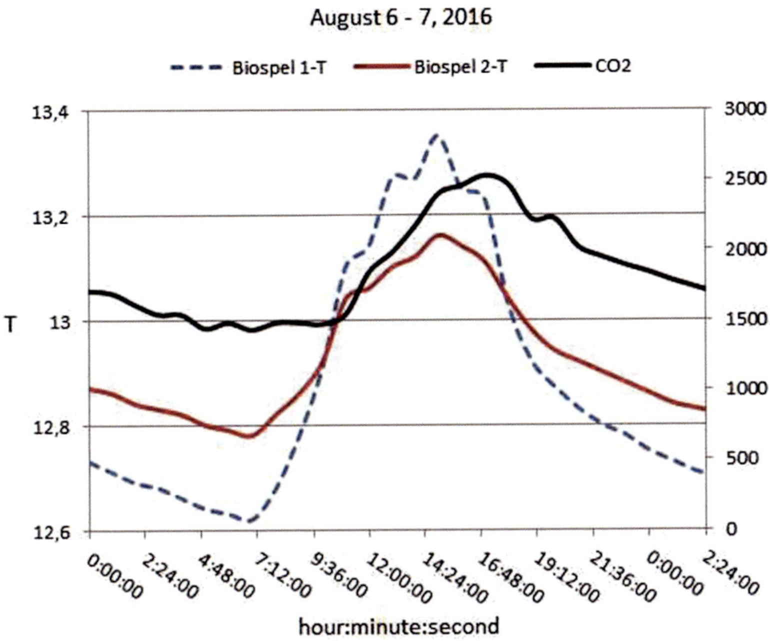

Figure 5.1.2.18 Daily CO_2 concentrations (ppm) and air temperature (°C) oscillations in Biospeleološka Postaja (August 6 to 7, 2016).

The differences between CO_2 concentrations in the morning (8:00 a.m.) and afternoon (17:00 p.m.) in July to August 2016 are presented in Fig. 5.1.2.19. Morning (previsiting) hours show constant values between 1200 and 1480 ppm. Afternoon values, which are strongly impacted by the breathing of visitors, are in the range of 1570—3130 ppm. The difference between the highest and lowest daily values is 250—1800 ppm, with a mean value of 750 ppm. This difference in CO_2 concentration is related to human presence in summer 2016, when metal doors were not yet opened full time.

The summer period (July 20 to August 7, 2019) highlights different outcomes (Fig. 5.1.2.20) to summer 2016. In 2019, metal doors were constantly opened and afternoon CO_2 concentrations were between 920 and 1890 ppm, while morning values were 1070—1560 ppm. Open metal doors caused afternoon concentrations to be more similar to the morning concentrations and can even cause them to be lower than morning concentrations with negative difference values (Fig. 5.1.2.20).

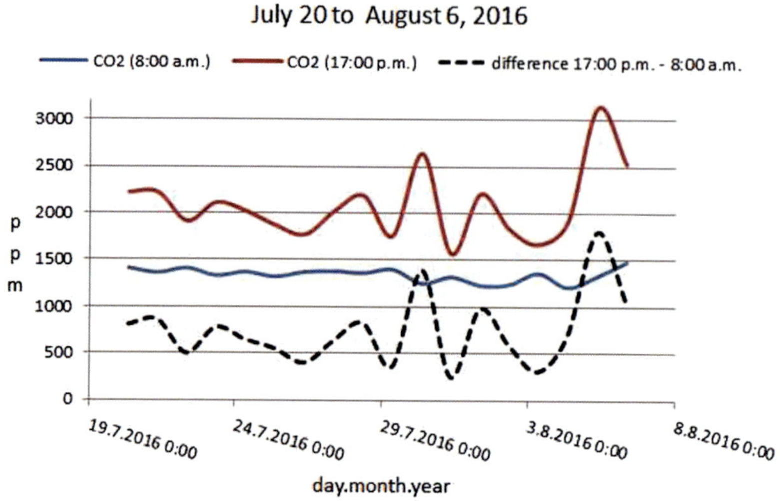

Figure 5.1.2.19 Morning (8:00 a.m.) and afternoon (17:00 p.m.) CO_2 concentrations (ppm) in Biospeleološka Postaja (July 20 to August 6, 2016).

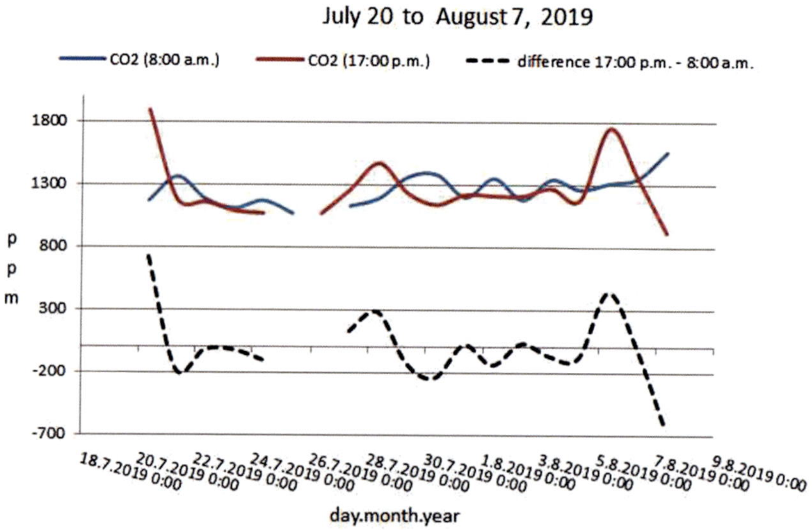

Figure 5.1.2.20 Morning (8:00 a.m.) and afternoon (17:00 p.m.) CO_2 concentrations (ppm) in Biospeleološka Postaja (July 20 to August 7, 2019).

Deposition of flowstone

Biospeleološka Postaja is also the location where, in 1963, well-known Slovene karstologist Dr. Ivan Gams (1923–2014) placed a glass plate (61 × 18 cm) under the area that water was percolating from the ceiling (Fig. 5.1.2.21). In the period 1963–98 it was calculated that an average of 0.11 mm of flowstone per year was deposited on the horizontal glass plate (Gams and Kogovšek, 1998). The flowstone accumulation mostly depends on the time of degassing of the percolating water between the ceiling and glass plate. In the period 1964–2004, 111 mg of flowstone was deposited per day, or 1690 mg over the entire period (Gams, 2004). The glass plate is covered by a very pure white flowstone, which is still depositing today. This shows that cave microclimatic conditions and outside precipitation remain favorable for flowstone deposition at this location.

Figure 5.1.2.21 Glass plate (61 x 18 cm) covered by flowstone is placed in Biospeleološka Postaja since 1963. *(Photo: S. Šebela.)*

CHAPTER 5.1.3

The most visited parts of Postojna Cave

The most visited parts of Postojna Cave include the areas with a cave train path (Pulpito–Veliki Dom and Sepolcro–Razpotje) and those with a walking path (Velika Gora, Lepe Jame and Koncertna Dvorana). Six monitoring sites are located close to the tourist path (1–15 m) (Fig. 5.1.3.1). The first air temperature monitoring site was established on July 21, 2008, at Postojna 2 in Lepe Jame passage (Šebela and Turk, 2011a, 2011b). On March 18, 2009, the second monitoring site at Velika Gora was created. An additional monitoring site at Lepe Jame (Postojna 3) was established on May 6, 2009. In the first period of measurements (July 21, 2008 to August 12, 2010) at the three locations (Velika Gora, Lepe Jame two and Lepe Jame 3) the air temperature data set included a number of periods with no measurements (Šebela and Turk, 2011b). The reasons for this were mistakes in organization of measurements, mistakes in data reading, and the absence of air temperature instruments. However, despite this, preliminary air temperature data showed good outlines of the beginning of modern systematic long-term microclimatic monitoring in Postojna Cave. Especially Postojna 2 site showed the highest air temperatures in winter and the lowest in summer (Šebela and Turk, 2011b), which was unexpected, bearing in mind that the monitoring site is adjacent to a well visited passage. Additionally, the annual air temperature range for the Postojna 2 site was between 9.7 and 11°C, and was higher than at the Postojna 3 and Velika Gora sites (Šebela and Turk, 2011b). The first studies pointed to the necessity for the continuation of microclimatic monitoring and expansion of monitoring sites to other parts of the cave system.

In Koncertna Dvorana, an air temperature instrument was placed on October 5, 2011. The intention was to monitor the cave atmosphere before, during, and after the construction of a new kiosk-shop. Air temperature monitoring there is still on-going.

Two additional air temperature monitoring sites have been established at historic sites (Crestani and Anelli, 1939). Sepolcro monitoring site has

Natural and Anthropogenic Impacts on Cave Climates
ISBN 978-0-12-822954-5
https://doi.org/10.1016/B978-0-12-822954-5.00009-3

Figure 5.1.3.1 Monitoring sites in the most visited parts of Postojna Cave.

been in place since March 19, 2015, and another at the Pulpito site since December 19, 2016.

Although it appear that constant, longer, air temperature monitoring at selected monitoring sites will not show dramatic differences between years due to a stable cave microclimate, this is untrue. The fact is that air temperature differences are not large, but they exist. If we have warmer winter than normal, then the cave air responds to such outside-cave conditions.

If we have a very cold winter with weeks-long outside air temperature below 0°C and with a strong wind, cave air is impacted differently than in a mild winter.

In addition to air temperature monitoring, periodical monitoring of carbon dioxide concentrations is organized at Lepe Jame passage at the Postojna 2 and 3 sites (Šebela et al., 2013; Prelovšek et al., 2018). The cave walking path that goes through Lepe Jame is frequented daily by visitors, representing the most visited area in Postojna Cave. This was the reason that carbon dioxide measurements were added to air temperature monitoring for long-term systematic monitoring of the cave microclimate to enable its sustainable use for tourism.

CHAPTER 5.1.3.1

Pulpito

One of the microclimatic monitoring sites in the most visited parts of Postojna Cave is Pulpito—Veliki Dom (Fig. 5.1.3.1 in Chapter 5.1.3, The most visited parts of Postojna Cave). This location was fixed on December 19, 2016, at the same location that the historic air temperature monitoring was set (Crestani and Anelli, 1939). The Pulpito site (Fig. 5.1.3.1.1) gives microclimatic information about the area between the Postojna Cave main entrance (see Chapter 5.1.1, Postojna Cave entrance part), Veliki Dom, and Biospeleološka Postaja (see Chapter 5.1.2, Biospeleological Station). The annual air temperature range at the Pulpito site is impacted by the vicinity of the cave entrance (see Fig. 5.1.2.1 in Chapter 5.1.2, Biospeleological Station). It was 13.34°C in 2017, 13.72°C in 2018, and 11.38°C in 2019. If we compare the annual air temperature range from the Pulpito site with Postojna Cave entrance (see Table 5.1.1.1 in Chapter 5.1.1, Postojna Cave

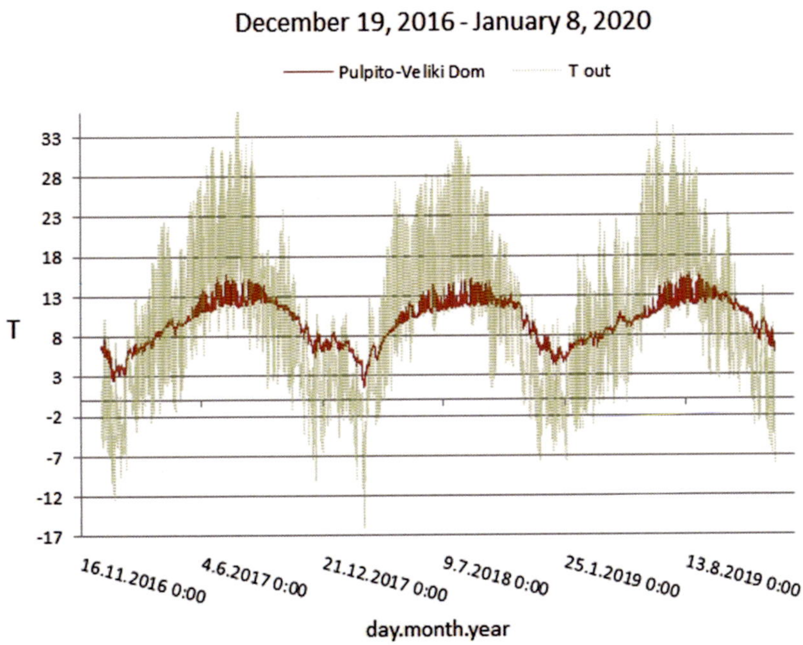

Figure 5.1.3.1.1 Hourly air temperature measurements at Pulpito monitoring site compared to measurements outside the cave in °C (December 19, 2016 to January 8, 2020).

Natural and Anthropogenic Impacts on Cave Climates
ISBN 978-0-12-822954-5
https://doi.org/10.1016/B978-0-12-822954-5.00010-X

Table 5.1.3.1.1 Basic statistics for air temperature (°C) at Pulpito and outside the cave for the period 2017–2019.

	Pulpito—Veliki Dom	Outside
2017		
Mean	9.74	10
Min	2.46	−12.31
Max	15.8	37.07
Range	13.34	49.38
2018		
Mean	9.73	10.55
Min	1.48	−15.94
Max	15.2	33
Range	13.72	48.94
2019		
Mean	10.09	10.58
Min	4.32	−7.51
Max	15.7	34.92
Range	11.38	42.43

entrance) we notice that the values are much lower at the Pulpito site. Annual air temperature range (January 7, 2013 to January 7, 2014) at Postojna Cave entrance was 29°C (at 6.2 m), 33.6°C (at 3.6 m), and 25.5°C (at 0.2 m). The Pulpito site represents the space between the cave entrance and inner parts.

Mean annual air temperature at the Pulpito site (Table 5.1.3.1.1) was 9.74°C in 2017, 9.73°C in 2018, and 10.09°C in 2019. The mild winter in 2019 caused a minimum air temperature value at Pulpito site of 4.32°C, while it was 1.48°C in 2018. This monitoring location shows the strong impact of outside meteorological conditions on winter cave air temperatures. The outside summer meteorological conditions do not show as strong an impact on summer cave air temperatures. Maximum air temperature values were 15.8°C in 2017, 15.2°C in 2018, and 15.7°C in 2019 (Table 5.1.3.1.1). This shows that winter ventilation ($T_{cave} > T_{out}$) has a greater impact on air temperature at the Pulpito site than summer ventilation ($T_{cave} < T_{out}$). More support for this is seen in Table 5.1.3.1.2. Based on hourly air temperature measurements, Pearson correlation coefficients (PCCs) showed a strong correlation between the Pulpito site and outside for the years 2017 (PCC = 0.72) and 2018 (PCC = 0.73). However, the PCC for 2019 (PCC = 0.63) presents a medium correlation between the

Table 5.1.3.1.2 Pearson correlation coefficients (PCC) for Pulpito and outside for hourly air temperature data.

	PCC
Pulpito—out (2017)	0.72
Pulpito—out (2018)	0.73
Pulpito—out (2019)	0.63

0.90—0.99 (very strong correlation)
0.70—0.89 (strong correlation)
0.40—0.69 (medium correlation)

Pulpito site and outside, in addition to the highest minimum value in 2019 and the lowest annual air temperature range in 2019. The mild winter of 2019 did not have a strong influence on Pulpito air temperature, as 2017 and 2018 had, resulting in a relatively weak correlation between the Pulpito site and outside on an annual level.

Three-year (2017—19) mean air temperature at the Pulpito site (9.85°C) was 0.52°C lower than the mean outside air temperature for the same period (Table 5.1.3.1.3). Here we discuss that although the Pulpito site is close to the underground River Pivka, the situation is different to Otoška Jama (see Chapter 5.1.5, Otoška Jama), where the mean annual air temperature is 1.95°C higher than outside. In Veliki Dom chamber and at the Pulpito site, the underground River Pivka does not have as strong an impact on cave air temperature as in the Otoška Jama 2 site. Winter ventilation in Veliki Dom and at Pulpito can strongly lower mean annual air temperature. Otoška Jama 2 site gave a result of 11.35°C in 2019 as the minimum winter value owing to the fact that this location does not have a large natural entrance and consequently the winter ventilation at Otoška Jama 2 is much weaker than at the Pulpito site.

It has been mentioned by Crestani and Anelli (1939) that the Pulpito air temperature data did not show any connection with the presence of people sitting on the train, which was passing close to the monitoring site. In 1939 (Crestani and Anelli) it was concluded that the air temperature at Pulpito is strongly dependent on the movement of air masses coming from Veliki

Table 5.1.3.1.3 Difference in °C between mean Pulpito cave air temperature and mean outside temperature for period 2017—2019.

	2017—2019	ΔT (Pulpito—outside)
Pulpito-Veliki Dom	9.85	−0.52
Outside	10.37	

Dom due to the differences between cave air and outside air. The findings published in 1939 (Crestani and Anelli) are in accordance with today's conditions (Fig. 5.1.3.1.2).

From July to October (Fig. 5.1.3.1.2), daily air temperature oscillations are the highest over a period of 1 year and can reach ~3°C. The maximum air temperature value at Pulpito was reached on September 4, 2019, at 3:00 a.m. (15.7°C), while outside it was 13.14°C at the same time. This means that cave air was flowing out of the cave because $T_{cave} > T_{out}$. The highest outside air temperature (34.92°C) was measured on June 26, 2019, at 14:00 p.m., while at the same time it was 10.87°C at Pulpito. The second highest outside air temperature peak was on July 25, 2019, at 12:00 p.m. At the same time, at the Pulpito site air temperature was 11.44°C. For both summer days when $T_{cave} < T_{out}$, warm outside air entered the cave through Veliki Dom. The air temperature peak for outside the cave does not overlap with the Pulpito air temperature peak. Pulpito has a lagged peak for about 1–2 months after the outside monitoring location. The lowest winter values at Pulpito show a strong dependence on the outside

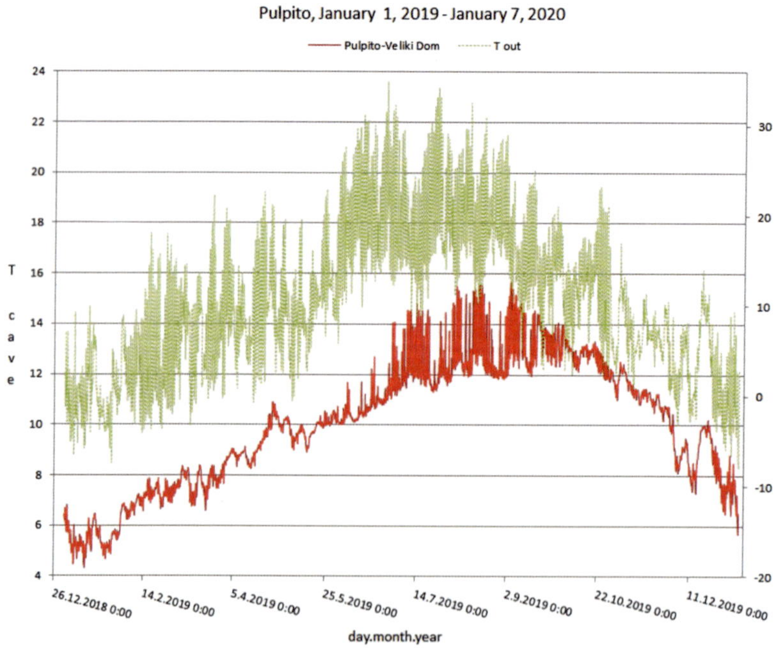

Figure 5.1.3.1.2 Hourly air temperature data at Pulpito site compared to outside air temperature in °C (January 1, 2019 to January 7, 2020).

climate, especially when outside air temperature is below 0°C. This was described in Chapter 5.1.1 (Fig. 5.1.1.9).

For the Pulpito site, black coatings on speleothemes on cave walls are typical, as described for Veliki Dom in Chapter 5.1.1 (Postojna Cave entrance).

CHAPTER 5.1.3.2

Velika Gora

Contents

Postojna 1 air temperature monitoring site sits 560 m above sea level on the top of the collapse blocks of the Velika Gora (Big Mountain) collapse chamber and is the highest monitoring point in the touristic part of Postojna Cave. Velika Gora is one of the largest chambers in Postojna Cave with a volume of 240,000 m^3 (Šebela, 1998). Air temperature monitoring has been on-going at hourly intervals since March 18, 2009. Cross-section AB in Fig. 5.1.3.2.1 shows the position of the air temperature instrument in Velika Gora chamber. The northern edge of the chamber follows the Dinaric-oriented (NW—SE) fault zone with morphological traces of reverse faulting (Fig. 5.1.3.2.2). This is the place where, in 2004, the TM71 extensometer monitoring of microdisplacements in Postojna Cave started (Šebela et al., 2010). Five years after setting up the extensometer monitoring, automatic air temperature monitoring was additionally organized at the same location (Fig. 5.1.3.2.3). Manual air temperature readings (Fig. 5.1.3.2.3) have been organized parallel with extensometer monitoring since 2004.

In order to understand year-round air temperature behavior at Velika Gora site we need to compare it with other close sites in the cave and also with outside conditions (Figs. 5.1.3.2.1 and 5.1.3.2.4). It is clear that hourly temperature oscillations at Velika Gora are lower than in Lepe Jame and in Koncertna Dvorana. Velika Gora monitoring site has a larger cave room space than the Lepe Jame and Koncertna Dvorana sites. Especially in summer, daily temperature oscillations are ~0.1°C at Velika Gora, while they are 0.38°C in Koncertna Dvorana. On August 24, 2019, at 15:00 p.m. at Velika Gora we measured a maximum air temperature of 11.9°C (Table 5.1.3.2.1), which is the same as the outside air temperature peak 1 month later.

The first change in the Velika Gora temperature curve, starting on April 20, 2019, is related to an increase in visits to the cave due to Easter—May 1st holidays. The second increase in air temperature at the end of 2019 was due

Natural and Anthropogenic Impacts on Cave Climates
ISBN 978-0-12-822954-5
https://doi.org/10.1016/B978-0-12-822954-5.00011-1

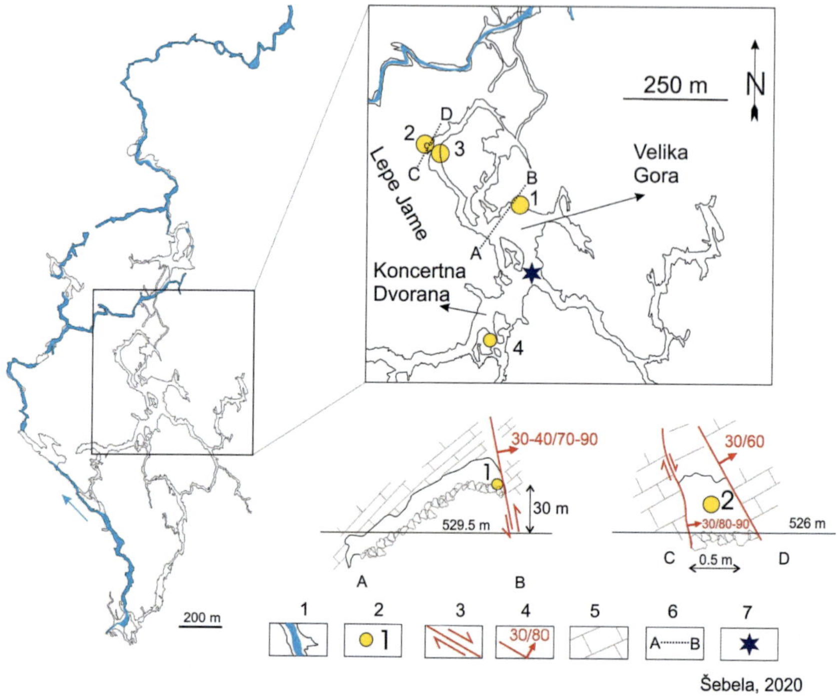

Figure 5.1.3.2.1 Postojna 1 (Velika Gora) temperature monitoring site together with other monitoring sites in the vicinity. Legend: 1 — underground River Pivka, 2 — air temperature monitoring sites (1 - Postojna 1 (Velika Gora), 2 - Postojna 2 (Lepe Jame 2), 3 - Postojna 3 (Lepe Jame 3), 4 - Koncertna Dvorana), 3 — vertical movement along the fault, 4 — geological elements of fault strike/dip in degrees, 5 — limestone bedding, 6 — cross-sections AB and CD, 7 — historic air temperature monitoring site (Schmidl 1854).

Figure 5.1.3.2.2 Velika Gora chamber, view from air temperature monitoring site Postojna 1 towards SE with tectonic displacement direction of the fault zone. *(Photo: S. Šebela.)*

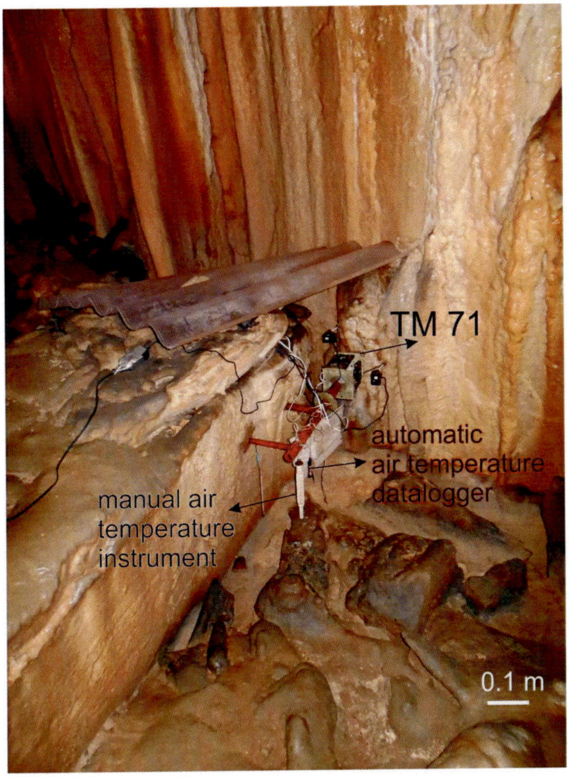

Figure 5.1.3.2.3 Velika Gora air temperature and micro-displacement (TM71) monitoring site. *(Photo: S. Šebela.)*

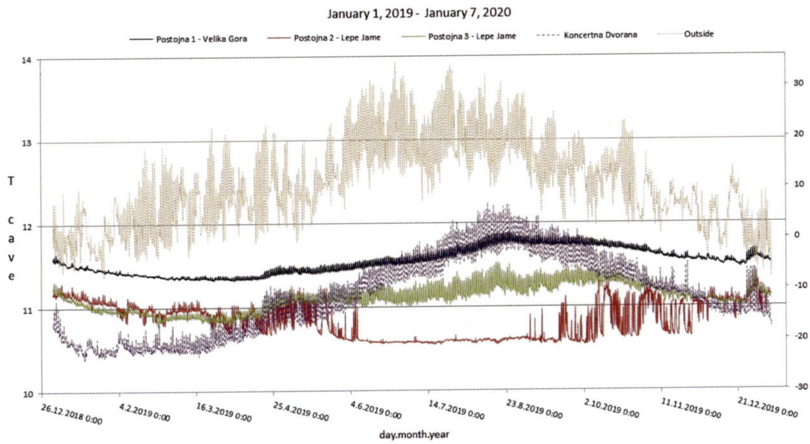

Figure 5.1.3.2.4 Air temperature data set in °C at hourly intervals for Velika Gora, Lepe Jame, Koncertna Dvorana and outside the cave sites for January 1, 2019 to January 7, 2020.

Table 5.1.3.2.1 Basic statistics for Velika Gora, Lepe Jame, Koncertna Dvorana and outside for 2019 in °C.

	Postojna 1—Velika Gora	Postojna 2—Lepe Jame	Postojna 3—Lepe Jame	Koncertna Dvorana	Outside
Mean	11.54	10.83	11.1	11.11	10.58
Max	11.9	11.45	11.53	12.24	34.92
Min	11.32	10.54	10.76	10.37	−7.51
Range	0.58	0.91	0.77	1.87	42.43

to increased visits because of Christmas performances, which traditionally take place in the near vicinity of Velika Gora monitoring site.

The annual air temperature range at Velika Gora is 0.58°C (Table 5.1.3.2.1), which is considerably lower than at the Lepe Jame and Koncertna Dvorana sites. In fact, this value is the lowest annual range over all the temperature monitoring sites in Postojna Cave, as presented in this study.

The Pearson correlation coefficient (PCC) for Velika Gora and outside the cave (Table 5.1.3.2.2) for a year-long period shows a medium correlation (PCC = 0.44), which has provoked a lot of thought as to why. The first reason must be the 1-month time-lag between the highest air temperature summer peak at Velika Gora compared with the outside situation. If we run a PCC calculation without time-lag we get PCC = 0.75, which gives a much stronger correlation between Velika Gora and outside on an annual air temperature data level.

Another question is why the air temperature at Velika Gora is constantly higher than at Lepe Jame (Postojna 2 and 3 sites). On an annual range, Velika Gora has ~0.5°C higher air temperature than the Postojna 3 site and 0.5−1.2°C higher values than Postojna 2. To understand this situation we must go back to the period between January and March 2012, when an interesting

Table 5.1.3.2.2 Pearson correlation coefficients (PCC) for Velika Gora, Postojna 2, Postojna 3 and Koncertna Dvorana and outside for hourly air temperature data January 1, 2019 to January 7, 2020.

	PCC
Velika Gora—outside	0.44
Postojna 2—outside	−0.72
Postojna 3—outside	0.45
Koncertna Dvorana—outside	0.72

0.90−0.99 (very strong correlation)
0.70−0.89 (strong correlation)
0.40−0.69 (medium correlation)

situation was described (Šebela et al., 2015a). Between more days long period with a strong northern wind called "bora" quick and significant temperature changes, attributed to fluxes of cold outside air penetrating deep into the cave, were observed at the Postojna 3 site in the Lepe Jame passage. At the Lepe Jame 3 site a temperature decrease of 0.3°C was observed on February 10, 2012, due to the bora wind (Šebela et al., 2015a). This situation was described as one of the most significant and rapid drops of air temperature at the Postojna 3 location created by natural drivers (Šebela et al., 2015a).

However, the situation was not the same at the Velika Gora site, where on February 10, 2012, there was no temperature decrease comparable to that at Lepe Jame (see Fig. 5.1.3.3.5 in Chapter 5.1.3.3, Lepe Jame), but the cave air temperature increased instead (Šebela et al., 2015a). This was described by the principle of stratification of cave air (with colder air near the floor and warmer air near the ceiling), which plays an important role in the thermal dynamics of the cave environment (Šebela et al., 2015a, b).

The increase in air temperature at Velika Gora was ~0.15°C. A drop in the air temperature for 0.3°C in Lepe Jame (Postojna 3 site) due to the "bora" wind, caused warmer cave air, which is not as heavy as colder air, to be "pushed" to higher elevations in the cave, in this case up to the highest levels in Velika Gora where the air temperature data logger is places. Here, it should be added that the top of Velika Gora acts as a "warm air trap" (Fig. 5.1.3.2.5), accumulating warm air even in times of strong intrusion of cold outside air deep in the cave, as in the winter of 2012. The orientation of the schematic profile AB of Velika Gora is the same as in Fig. 5.1.3.2.1. The winter and summer characteristics of a "warm air trap" on the top of the Velika Gora collapse chamber gives sufficient explication for the year-round increased air temperature values at this monitoring site compared with other sites in the area.

If the cave is a warm air trap it can be suitable as a summer roost for bats (Elliotte, 2012), but this is not the case for Velika Gora. Warm air traps are described as convection cells frequently occurring near large entrances (Covington and Perne, 2015). The relative elevation of the entrance and cave void determines whether convection is active in the winter or summer (Covington and Perne, 2015). In this case, Velika Gora site is not situated close to the cave entrance.

Due to the increased air temperature values at Velika Gora compared with other sites in the area, we can also expect a large difference between the mean cave air temperature and outside air temperature (Table 5.1.3.2.3).

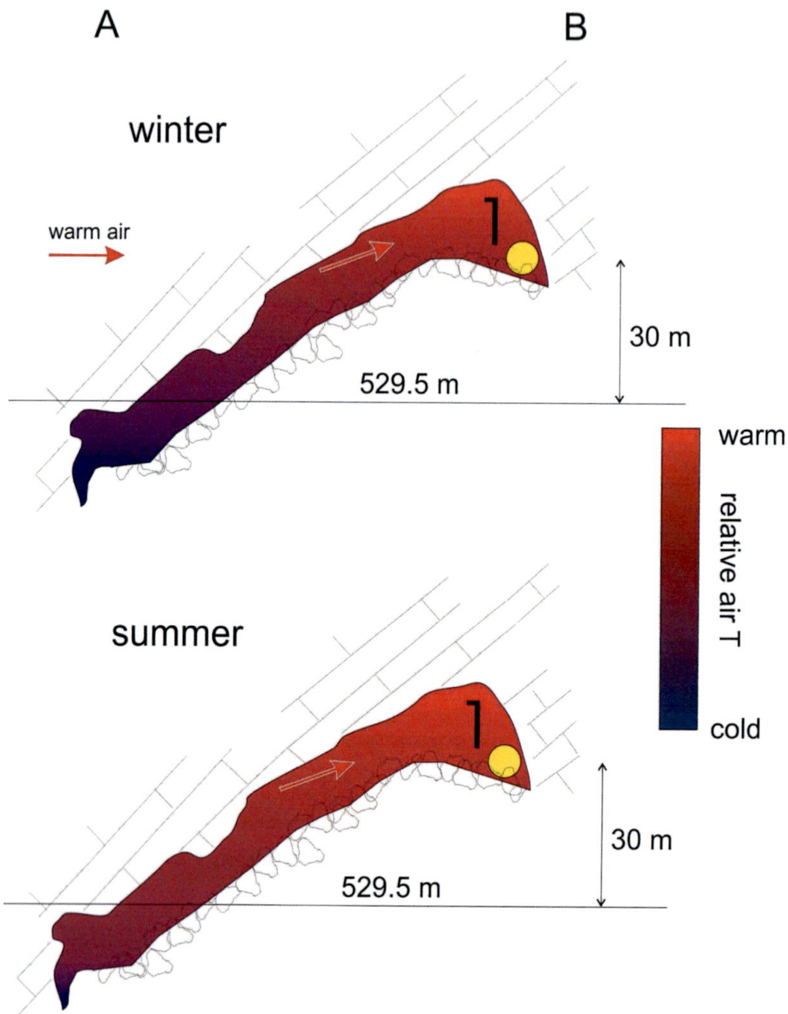

Figure 5.1.3.2.5 Cross section AB accross Velika Gora air temperature monitoring site (number 1) as "warm air trap" in winter and summer conditions.

Table 5.1.3.2.3 Difference between mean air temperature at Velika Gora and outside (2010–2019) in °C.

	Mean 2010–2019	ΔT (cave – surface)
Velika Gora (Postojna 1)	11.28	+1.39
Outside	9.89	–

For the period 2010–19, the mean air temperature at the Velika Gora site is 1.39°C higher than outside the cave. Only in Otoška Jama and Biospeleološka Postaja are there higher values.

At Velika Gora (Postojna 1 site) and in Lepe Jame (Postojna 2 site) the first air temperature measurements started in 2004 with monthly manual readings (Fig. 5.1.3.2.3). Even if such instruments do not have high precision, analyses of data sets and comparisons with automatic measurements can give additional information. First, we see that all curves are increasing (Fig. 5.1.3.2.6). An air temperature increase from automatic instruments is even higher than from manual thermometers, which show 0.43–0.47°C lower values than automatic instruments (Table 5.1.3.2.4). In the period 2012–15, the automatic

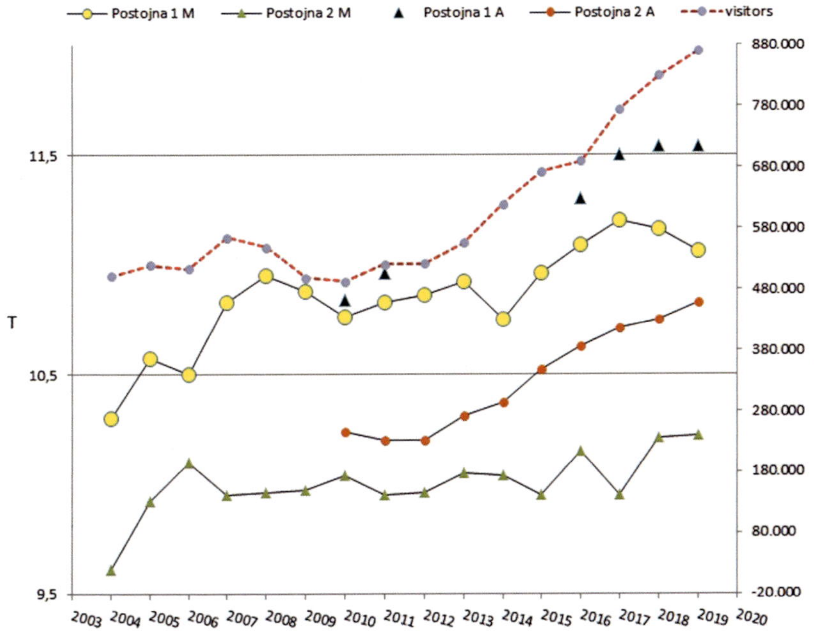

Figure 5.1.3.2.6 Mean annual air temperature at Postojna 1 and 2 sites (M – manual and A – automatic readings) compared with number of visitors per year in °C.

Table 5.1.3.2.4 Comparison between mean air temperature obtained by manual and automatic readings in C.

	Postojna 1	Postojna 2	
Mean (2004–2019)	10.85	10.00	M
Mean (2010–2019)	11.28	10.47	A
Difference (A–M)	0.43	0.47	

data logger was not installed at the Velika Gora site. The maximum mean annual value determined from manual readings at Velika Gora was 11.2°C in 2017. The minimum mean annual air temperature according to manual readings was 10.3°C in 2004. An automatic air temperature instrument at Velika Gora showed 11.54°C as the maximum mean annual value in 2019 and 10.84°C as the minimum value in 2010.

The PCC between manually measured air temperature data at Velika Gora and the number of visitors per year was 0.71, which is a strong correlation.

Regarding automatic hourly measurements, it can be seen that the mean annual air temperature at Velika Gora has been increasing since 2010, and additionally according to the manual monthly readings, the mean annual air temperature at Velika Gora has been increasing since 2004.

Historic air temperature measurements at Velika Gora

The historic measurements at Velika Gora should not be ignored. In Fig. 5.1.3.2.1 number 7 (blue star) represents the location of air temperature measurements performed by Schmidl (1854). From September 14–15, 1852, Schmidl (1854) performed 24 h air temperature measurements at floor level in the Velika Gora chamber. This was 14 years before the artificial enlargement and opening of the main Postojna Cave entrance in 1866, which means that the outside air currents did not have such easy access to the cave as they do today (Šebela et al., 2015a).

Schmidl (1854) determined the mean daily cave air temperature as being 8.5°C. After 9:00 a.m. on September 15, 1852, Schmidl (1854, page 169) described a temperature increase of 0.25°C due to the presence of a group of visitors with torches. There are no data regarding the number of visitors involved, but it probably did not exceed a few tens at the most. From 17:00 p.m. to 24:00 p.m. on September 15, 1852, the cave air temperature was 8.75°C. The increase in the air temperature can be attributed to the cumulative effect after visitation, presumably due to the poor ventilation in the cave before opening of the main cave entrance in 1866 (Šebela et al., 2015a).

The increase in the cave air temperature in Lepe Jame during feast days in recent years also reaches a maximum of 0.25°C (Šebela et al., 2013), however the number of visitors is now much higher than in 1852. Moreover, the mean cave air temperature now appears to be almost 2°C higher than in 1852 (Šebela et al., 2015a).

From February 20 to June 30, 1933, 14 air temperature measurements were performed by Crestani and Anelli (1939) in almost the same location as Schmidl's measurements in 1852. The mean value was 8.58°C, which is basically the same as in 1852 (Schmidl, 1854).

Historic measurements of cave air temperature were probably not performed with very precise instruments, however the same values were measured in 1852 and 1933, showing that the air temperature at the bottom of the Velika Gora collapse chamber did not significantly change over the period from 1852 to 1933. Taking into account that in 1866 the main entrance to Postojna Cave was artificially opened, it is interesting to see that in February–June 1933, the cave entrance modification did not impact the cave air temperature any differently to in September 1852.

CHAPTER 5.1.3.3

Lepe Jame (Postojna 3 site)

Contents

The location of Lepe Jame—Postojna 3 monitoring site is shown in Fig. 5.1.3.2.1 in Chapter 5.1.3.2 about Velika Gora. This site is only about 4 m from the tourist path (Fig. 5.1.3.3.1), which was used by 870,000 visitors in 2019. Regular air temperature monitoring at hourly intervals and periodical CO_2 measurements have been taking place here since May 6, 2009.

Air temperature monitoring

A graph showing the air temperature at Postojna 3 site in relation to other monitoring sites in the vicinity (Lepe Jame—Postojna 2, Koncertna Dvorana, and Velika Gora) is presented in Chapter 5.1.3.2 about Velika Gora (Fig. 5.1.3.2.4). Here, we present year-long hourly air temperature data from the Postojna 3 site and outside the cave (Fig. 5.1.3.3.2). In the colder part of the year, the 2019 daily temperature oscillations were less than 0.15°C, while in the

Figure 5.1.3.3.1 Lepe Jame – Postojna 3 air temperature and CO_2 monitoring site. *(Photo: S. Šebela.)*

Natural and Anthropogenic Impacts on Cave Climates
ISBN 978-0-12-822954-5
https://doi.org/10.1016/B978-0-12-822954-5.00012-3

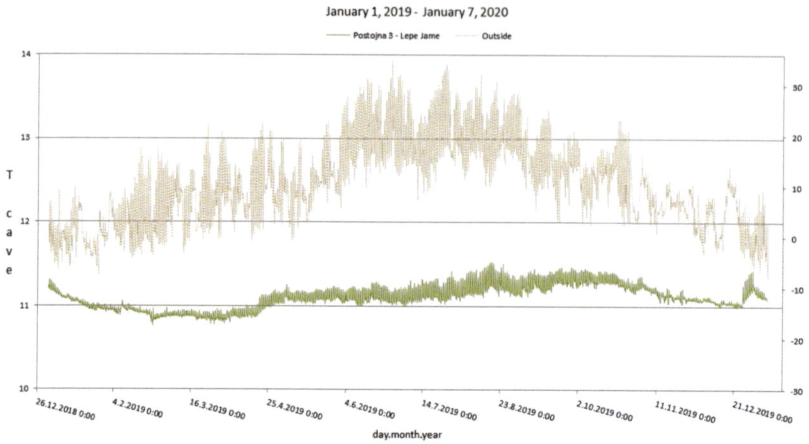

Figure 5.1.3.3.2 Hourly air temperature data set for Postojna 3 site in Lepe Jame in °C for the period January 1, 2019 to January 7, 2020.

warmer part they could reach 0.4°C. Annual air temperature oscillations are up to 0.77°C (see Table 5.1.3.2.1 in Chapter 5.1.3.2, Velika Gora). On April 20, 2019, we detected the first significant change in the temperature curve, which was caused by the increased number of visitors due to the Easter—May 1st holidays. On August 18, 2019, there was another temperature peak, also related to an increased number of visitors.

The autumn season from September to October 20, 2019 (Fig. 5.1.3.3.3), was characterized by another gentle temperature peak, which was related to the outside cave conditions. When we have a situation that $T_{out} > T_{cave}$, as from September to October 20, 2019, Lepe Jame was still under a summer ventilation regime, which caused a temperature peak in autumn. After October 20, 201,9 Lepe Jame switched to a winter ventilation regime with $T_{out} < T_{cave}$, causing a drop in the air temperature of ~0.2°C at Lepe Jame 3 site due to the intrusion of colder outside air into the cave.

The fourth air temperature peak started on December 25, 2019 (Fig. 5.1.3.3.4), with a 6-day Christmas performance event in Postojna Cave, when ~20,000 visitors visited the cave. Over the 6 days, the cumulative anthropogenic impact was an increased air temperature of ~0.2°C because after the performance event air temperature did not return completely back to the values of before December 25, 2019. Daily temperature oscillations were slightly higher. Approximately 3300 visitors per day increased the air temperature by 0.3°C in 1 day. Such a situation was described in previous years also (Šebela and Turk 2011a; Šebela and Turk 2014a; Šebela et al., 2015a).

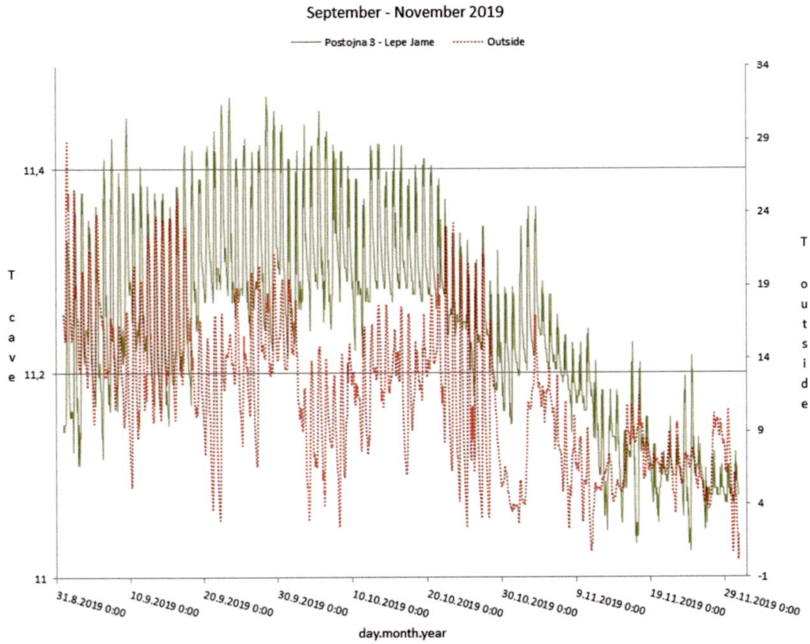

Figure 5.1.3.3.3 Postojna 3 — Lepe Jame hourly air temperature measurements September—November 2019 in °C.

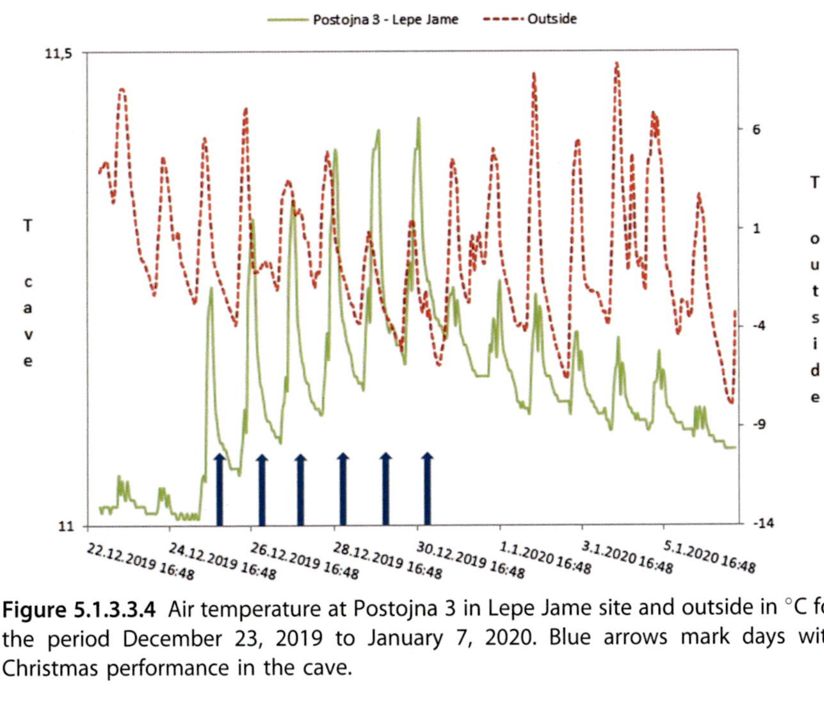

Figure 5.1.3.3.4 Air temperature at Postojna 3 in Lepe Jame site and outside in °C for the period December 23, 2019 to January 7, 2020. Blue arrows mark days with Christmas performance in the cave.

Table 5.1.3.2.1 in Chapter 5.1.3.2 (Velika Gora) gives basic statistical information about the Postojna 3 air temperature. The mean annual air temperature was 11.10°C in 2019, with a minimum value of 10.76°C and maximum of 11.53°C. Table 5.1.3.3.1 demonstrates by how much the mean air temperature at the Postojna 3 site was higher than the mean outside air temperature for the period 2010–19. This difference is +0.78°C, and is not as high as in Velika Gora, Otoška Jama, or Biospeleološka Postaja. The difference between the mean air temperature at Postojna 3 in 2010 compared with 2019 shows a +0.57°C higher mean air temperature in 2019 than in 2010.

An important situation at the Postojna 3 site happened in winter 2012 (Šebela et al., 2015a). This event took place between February 1–12 (Fig. 5.1.3.3.5), and was a consequence of blowing of the strong northern (bora) wind outside the cave. The mean wind speed was 3.8 m/s, with a

Table 5.1.3.3.1 Difference between mean surface and Postojna 3 air temperature for the period 2010–2019 in °C.

	Mean 2010–2019	**Δ*T* (surface − cave)**
Postojna 3	10.67	+0.78
Outside	9.89	

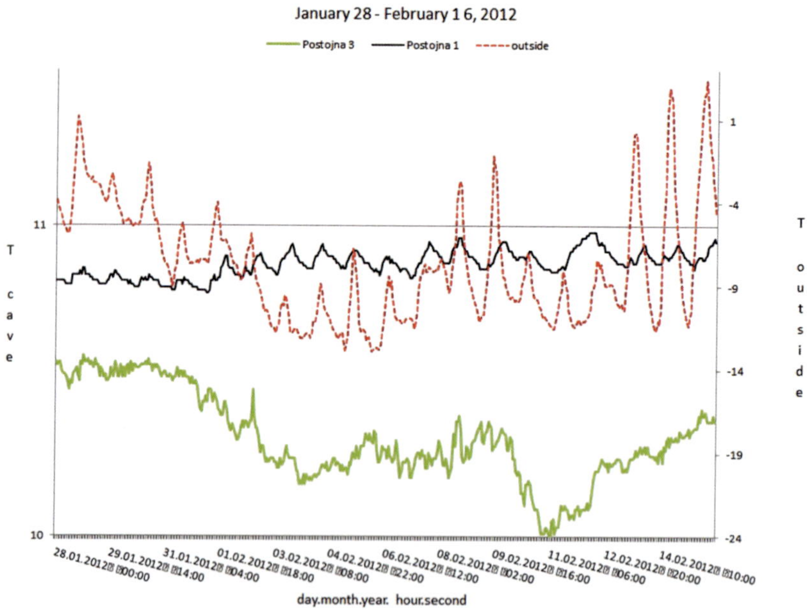

Figure 5.1.3.3.5 Cold natural event at Postojna 3 site, January 28 to February 15, 2012 in °C.

maximum speed of 18 m/s (Šebela et al., 2015a). It was observed that when a strong wind bora was blowing, cold air penetrated deep inside the cave (Postojna 3 site is ~1500 m from the main cave entrance). This caused a temperature decrease of 0.3°C to be observed at the Postojna 3 site in Lepe Jame on February 10, 2012 (Šebela et al., 2015a). Looking at a wider time window of February 1–12, 2012 (Fig. 5.1.3.3.5), the drop in air temperature was 0.5°C. A minimum air temperature of 9.99°C over the entire year 2012 was measured at the Postojna 3 site on February 11, 2012, while in 2019 the minimum air temperature was 10.76°C. The air temperature range at the Postojna 3 site in 2012 was 1.28°C, while it was 0.77°C in 2019.

This was one of the most significant and rapid drops in air temperature at this monitoring location between 2009 and 2013 (Šebela et al., 2015a), and even after 2013 we did not observe a similar situation repeated. Here, it should be added that natural driving factors such as extreme external climate events (for example, bora wind) can strongly contribute to fluxes of cold air entering cave passages (Šebela et al., 2015a). On the other hand, it must not be forgotten that during increased tourist visits at the Postojna 3 site (Fig. 5.1.3.3.4) we observed an air temperature increase of 0.3°C in 1 day. Extreme natural conditions can have an even greater impact on cave air temperature, such as anthropogenic effects due to increased visitor numbers.

About 20 m to the southeast of Postojna 3 monitoring site, the water temperature of the rimstone pool (Fig. 5.1.3.3.6) was compared with the air temperature at Postojna 3 site and the number of visitors for the monitoring period 2010–11 (Šebela and Turk, 2014a). The results showed that attention

Figure 5.1.3.3.6 Rimstone pool near Postojna 3 site in Lepe Jame, where water temperature was monitored at hourly intervals in the period 2010–2011 (Šebela and Turk 2014). *(Photo: S. Šebela.)*

must be paid to temperature changes of the rimstone pool water, especially during winter (Šebela and Turk, 2014a). It also was shown that anthropogenic effects had a relatively great impact on the water temperature in the Lepe Jame rimstone pool, especially in winter. The susceptibility of water temperature to anthropogenic influences was far greater than that of cave air. In Postojna Cave, heat absorption by pools and the rock mass was found to be relatively greater than in the cave air (Šebela and Turk, 2014a).

The Pearson correlation coefficient (PCC) for the Postojna 3 site and outside the cave (Table 5.1.3.2.2 in Chapter 5.1.3.2, Velika Gora) for 2019 shows a medium correlation (PCC = 0.45), similar to that at Velika Gora. Also, at the Postojna 3 site there is a time-lag of ~25 days between the highest outside temperature peak at the end of July 2019, and the highest air temperature peak in the middle of August 2019.

Carbon dioxide monitoring

At the Postojna 3 site, carbon dioxide (Fig. 5.1.3.3.1) hourly concentrations were measured over the course of six holiday periods in 2009–2011 (Šebela et al., 2013). A fivefold increase in visitor numbers during the Christmas performance event resulted in a carbon dioxide concentration that was at least twice as high as normal. The carbon dioxide concentration increased from 450 to 1750 ppm (December 2009–January 2010). After 10 days of peak period visits, both the CO_2 concentration and the temperature returned almost to preholiday values (Šebela et al., 2013).

In the next study (Prelovšek et al., 2018), periodical carbon dioxide hourly measurements covered not only holiday periods but also periods that were not overloaded with tourists from November 2012 to October 2015. A year-round CO_2 "distance-weighted" curve for the Lepe Jame site compared with the temperature and number of visitors per month showed two CO_2 peaks, the first in April–May and the second in September–October (Prelovšek et al., 2018). During spring and autumn, the daily alternation of the ventilation regime with a smaller rate of air exchange resulted in yearly cave air CO_2 peaks of up to ~2400 ppm (Prelovšek et al., 2018).

In this study we present two periods of CO_2 monitoring, which have not been published previously. The first period is from April 20–27, 2018 (Fig. 5.1.3.3.7). This is the spring period with an increased number of visitors due to national feast days, when the cave receives more than 3000 visitors per day. Comparing outside air temperature with cave air temperature we can see that Lepe Jame sites are subjected to summer ventilation during day hours ($T_{out} > T_{cave}$) and to winter ventilation

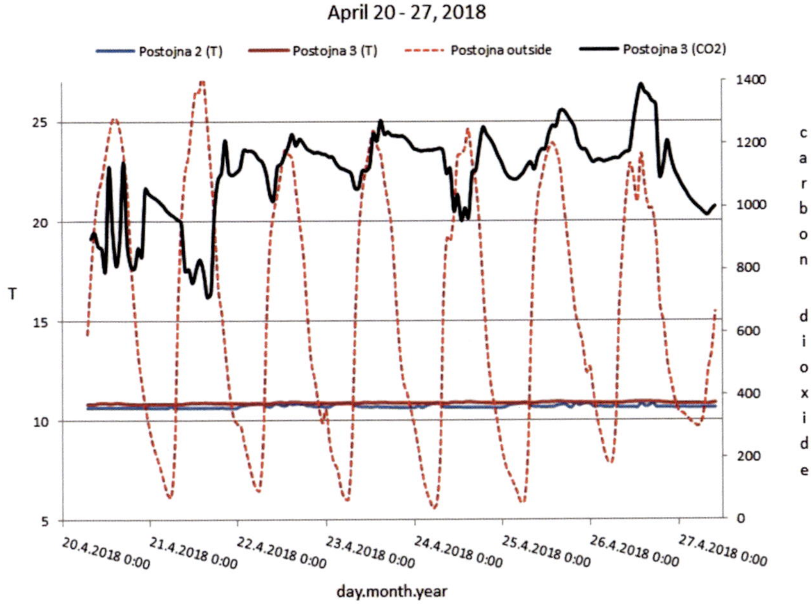

Figure 5.1.3.3.7 Carbon dioxide concentrations (ppm) at Postojna 3 site, outside cave and cave air temperature in °C for the period April 20–27, 2018.

($T_{out} < T_{cave}$) during night hours. There is a switch between the two ventilation regimes on a daily basis. Carbon dioxide levels were highest on April 26, 2018, when it was just a little below 1400 ppm at 14:00 p.m. On April 27, 2018, at 7:00 a.m., the carbon dioxide concentrations dropped to 980 ppm (Fig. 5.1.3.3.8). It is also noticed that the carbon dioxide concentration increased from April 20–27, 2018. This increase was about 400–500 ppm. Carbon dioxide concentrations during the May 1st feast days can be even higher. In the first 3 days of May 2014, they reached 2400 ppm (Prelovšek et al., 2018). The cause of an increase in carbon dioxide concentrations during the day hours is attributed to the increased visitor numbers on feast days. The level of carbon dioxide during night hours, at around 1000–1100 ppm, is attributed to natural conditions regarding winter ventilation regime and vegetation activity above the cave. Another interesting aspect is the remarkable day and night air temperature oscillations at Postojna 2 site, which is described in Chapter 5.1.3.4 [Lepe Jame (Postojna 2 site)].

The second period comprises the days from December 10, 2018, to January 8, 2019, and gives information about air temperature and carbon

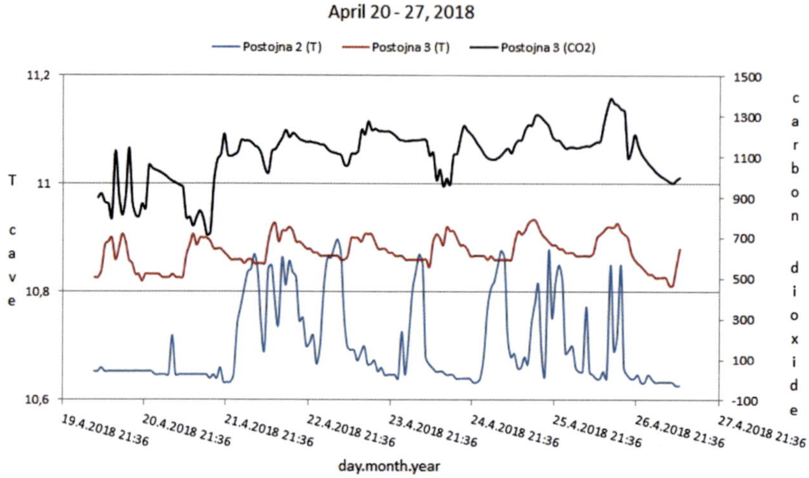

Figure 5.1.3.3.8 Carbon dioxide concentrations (ppm) at Postojna 3 site and cave air temperature in °C for the period April 20–27, 2018.

dioxide behavior during a period without high visitor numbers in the beginning and middle of December, and the period of the Christmas performance in the cave with high visitor numbers. Winter carbon dioxide concentrations at Lepe Jame 3 site before the Christmas performance days (December 25–30, 2018) were between 450 and 700 ppm (Fig. 5.1.3.3.9). Such values represent the natural winter background of cave CO_2, which

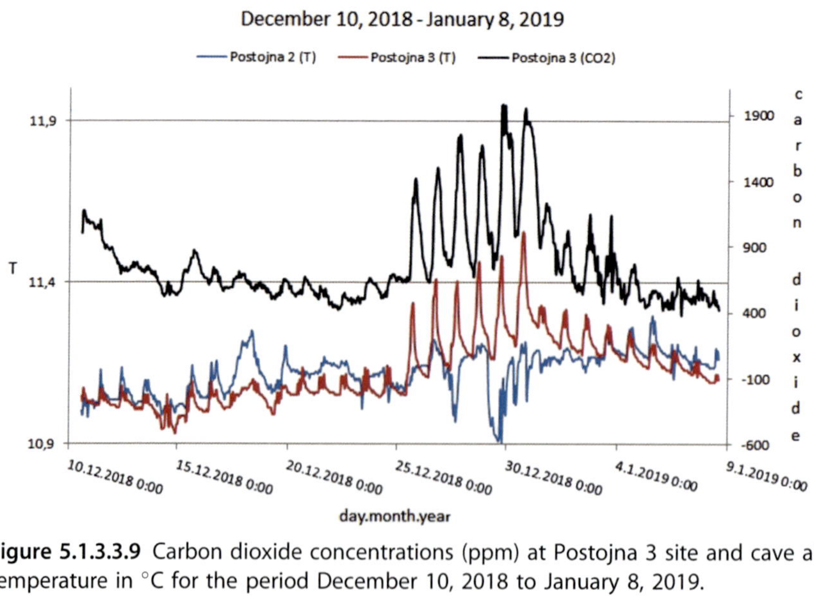

Figure 5.1.3.3.9 Carbon dioxide concentrations (ppm) at Postojna 3 site and cave air temperature in °C for the period December 10, 2018 to January 8, 2019.

does not show a strong impact of anthropogenic CO_2 due to the breathing of visitors. On feast days such as December 29, 2018, the maximum carbon dioxide concentration was 1970 ppm at 17:00 p.m. (Fig. 5.1.3.3.9), while the next day (December 30, 2018), the CO_2 concentration dropped to 1000 ppm at 9:00 p.m., which is close to the level before the visitors entered the cave. During the 6-day event the cave welcomed about 18,000 visitors, at about 3000 per day. After this event the CO_2 levels dropped quickly to the level of the concentrations before December 25, 2018. This was not the case with the air temperature at Postojna 3 site, which needed more days to recover to the level before the high visitor numbers.

CHAPTER 5.1.3.4

Lepe Jame (Postojna 2 site)

Contents

The Postojna 2 air temperature and carbon dioxide monitoring site (Fig. 5.1.3.4.1) is placed in a side passage near a microdisplacement monitoring site with a TM71 extensometer (Šebela et al., 2010). There is an air distance of about 20–30 m between the Postojna 2 and 3 monitoring sites (see Fig. 5.1.3.2.1 in Chapter 5.1.3.2, Velika Gora). Postojna 2 site (Fig. 5.1.3.4.2) is only ~35 m to the northwest of the heavy visited Lepe Jame passage. In the 1960s, speleologists enlarged the cave passage because there was a strong ventilation coming from a crack behind the known passage. They discovered only about 10–15 m of new passage, because the end was too narrow for continuation. About 10 years ago, speleologists added ~25 m of passage continuation in a very narrow crack.

The first air temperature readings at the Postojna 2 site were taken with manual thermometers on May 18, 2004 (see Fig. 5.1.3.2.6 in Chapter 5.1.3.2, Velika Gora). The readings were generally performed once per month. Additional automatic measurements at the same location were started at hourly

Figure 5.1.3.4.1 Position between Postojna 2 and 3 monitoring sites in Lepe Jame, view towards west. *(Photo: S. Šebela.)*

Natural and Anthropogenic Impacts on Cave Climates
ISBN 978-0-12-822954-5
https://doi.org/10.1016/B978-0-12-822954-5.00013-5

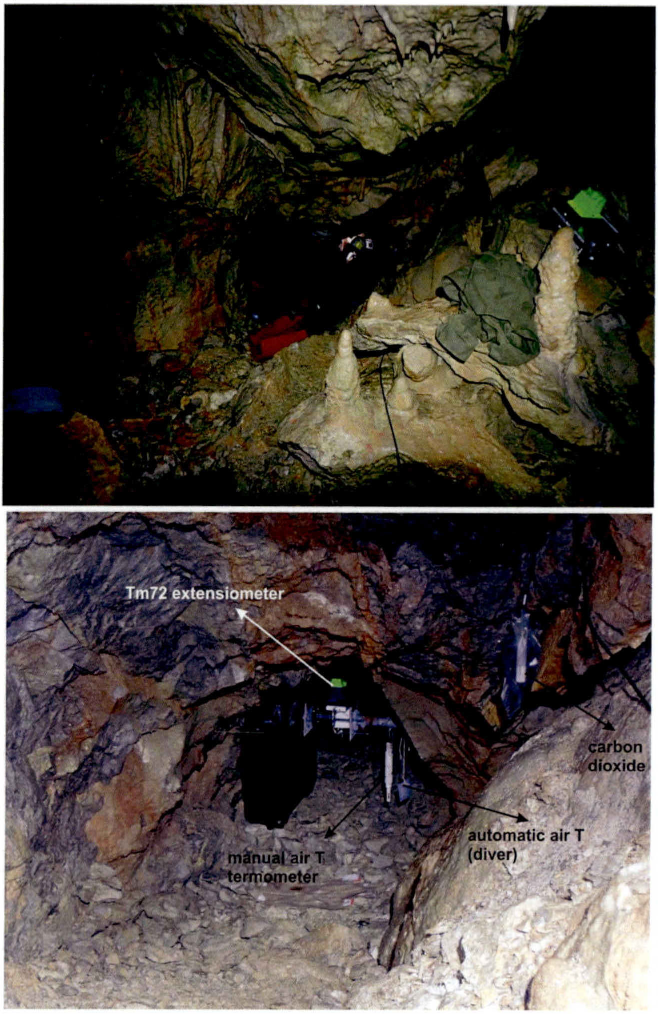

Figure 5.1.3.4.2 Narrow artificially enlarged passage at Postojna 2 monitoring site in Lepe Jame. *(Photo: S. Šebela.)*

intervals on July 21, 2008 (Šebela and Turk, 2011b). At the Postojna 2 site, standard deviation (SD) of air temperature for the period August 12, 2009, to August 12, 2010, was 0.27 and was much higher as at Postojna 1, Velika Gora (SD = 0.12), and the Postojna 3 site (SD = 0.09) (Šebela and Turk, 2011b). Systematic automatic air temperature monitoring at the Postojna 2 site showed the highest air temperature values during winter and the lowest during summer, which was explained by the air circulation from unknown passages behind the monitoring site (Šebela and Turk, 2011a). In January–March 2009, the data for snow cover and rain precipitation did not give a good correlation with temperature at Postojna 2, which indicated that this site is an interesting microclimate location (Šebela and Turk, 2011a).

A comparison of air pressure measured at 30 min intervals between the Postojna 2 site and outside the cave showed simultaneous pressure variations at the surface and underground, with no phase shift between cave pressure and external pressure (Šebela and Turk, 2011a). Based on this, it was presumed that the cave system is well ventilated, resulting in simultaneous exchanges of air pressure (Šebela and Turk, 2011a).

Air temperature monitoring

This monitoring site shows specific characteristics between all monitoring sites in Postojna Cave (Fig. 5.1.3.4.3). Air temperature behavior is opposite to outside the cave and other cave monitoring sites (Velika Gora, Postojna 3, Lepe Jame, Koncertna Dvorana). The lowest air temperature is during summer, while during the cold part of the year the air temperature at Postojna 2 is the highest. In summer there are almost no temperature oscillations, while in the colder part of the year daily oscillations can reach 0.7°C. In 2019, the air temperature range was 0.91°C (see Table 5.1.3.2.1 in Chapter 5.1.3.2, Velika Gora). Mean annual air temperature at the Postojna 2 site in 2019 was 10.83°C, which is 0.27°C lower than at the Postojna 3 site and 0.71°C lower than at the Velika Gora site. Especially in the summer period, the Postojna 2 site preserves low air temperature compared with other sites in the vicinity, such as the Postojna 3 and Postojna 1 sites.

Pearson correlation coefficient (PCC) for the Postojna 2 site and outside the cave (Table 5.1.3.4.1) for a year (2019) shows a strong negative or inversed correlation (PCC = −0.72). This is the only monitoring site in Postojna Cave with negative air temperature correlation compared to outside. The correlation between the Postojna 2 and 3 sites gives a very

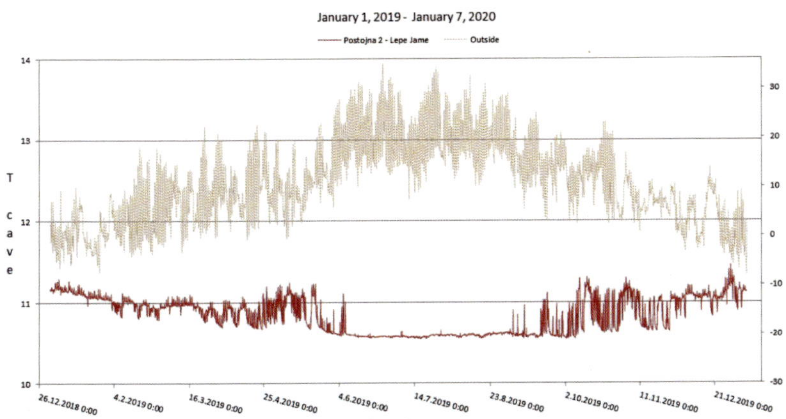

Figure 5.1.3.4.3 Postojna 2 air temperature data set compared with outside the cave conditions for January 1, 2019 to January 7, 2020 at hourly intervals in °C.

Table 5.1.3.4.1 Pearson correlation coefficients (PCC) for Postojna 2 site, Postojna 3 and outside for hourly air temperature data January 1, 2019—January 7, 2020.

	PCC
Postojna 2—outside	−0.72
Postojna 2—Postojna 3	−0.31

0.90—0.99 (very strong correlation)
0.70—0.89 (strong correlation)
0.40—0.69 (medium correlation)
0.20—0.39 (weak correlation)

weak negative correlation (PCC = −0.31) for the year-long period. Microclimate studies at Postojna 2 showed local variability, especially in winter, and exhibit special microclimate conditions that may be due to ventilation from unknown passages at the back (Šebela and Turk 2011a).

To understand the cause of a negative correlation we need to look closer at the periods with air temperature oscillations at the Postojna 2 site (Fig. 5.1.3.4.4). The Pearson correlation coefficient (PCC) for the Postojna 2 site and outside for the period May 15—June 12, 2019, gives a medium negative correlation (PCC = −0.45). This means that there is some negative dependence between the air temperature values at the Postojna 2 site and outside the cave, but this cannot be the main cause of air temperature changes at Postojna 2. If we show a shorter time period of May 18—23, 2019 (Fig. 5.1.3.4.5), the Pearson correlation coefficient (PCC) for the Postojna 2 site and outside the cave shows no dependence between the cave air temperature and outside cave conditions (PCC = 0.03).

To obtain more explanations regarding the behavior of air temperature at the Postojna 2 site we need information about the positions of known passages, the thickness of the limestone above the cave, and speleological investigations

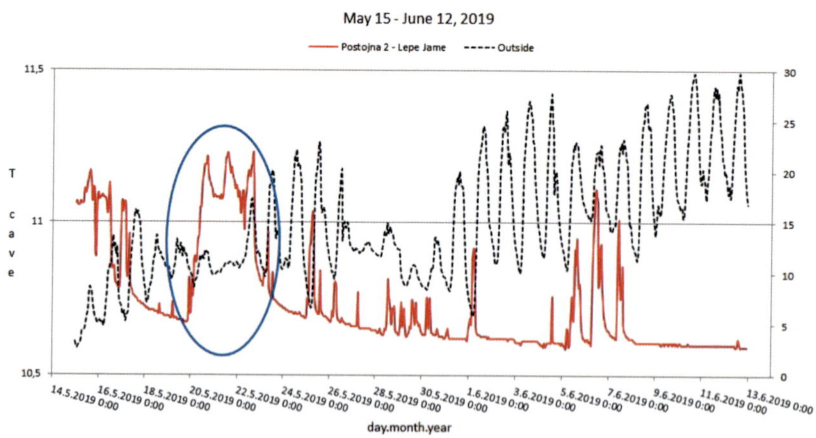

Figure 5.1.3.4.4 Relation between air temperature at Postojna 2 site and outside the cave in °C for the period May 15—June 12, 2019. Blue elipse is enlarged in Fig. 5.1.3.4.5.

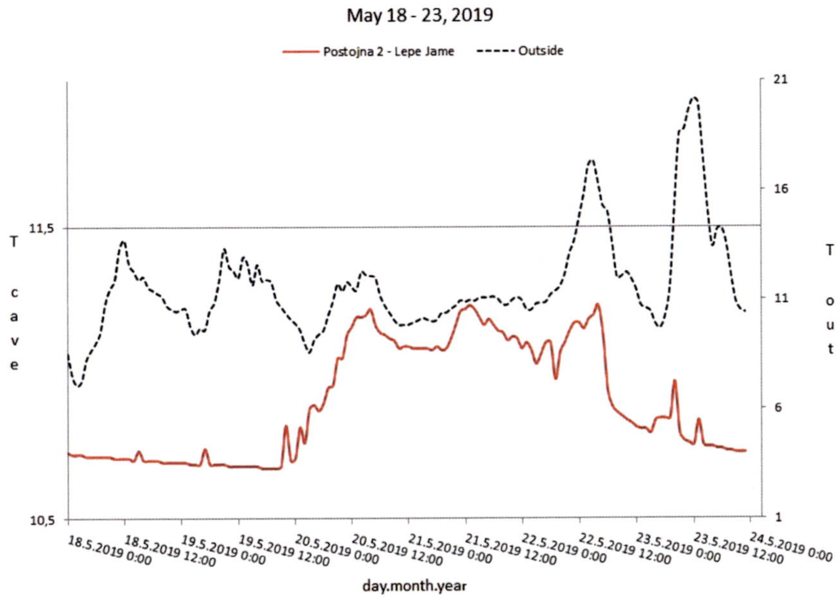

Figure 5.1.3.4.5 Relation between air temperature values at Postojna 2 site and outside the cave in °C for the period May 18–23, 2019.

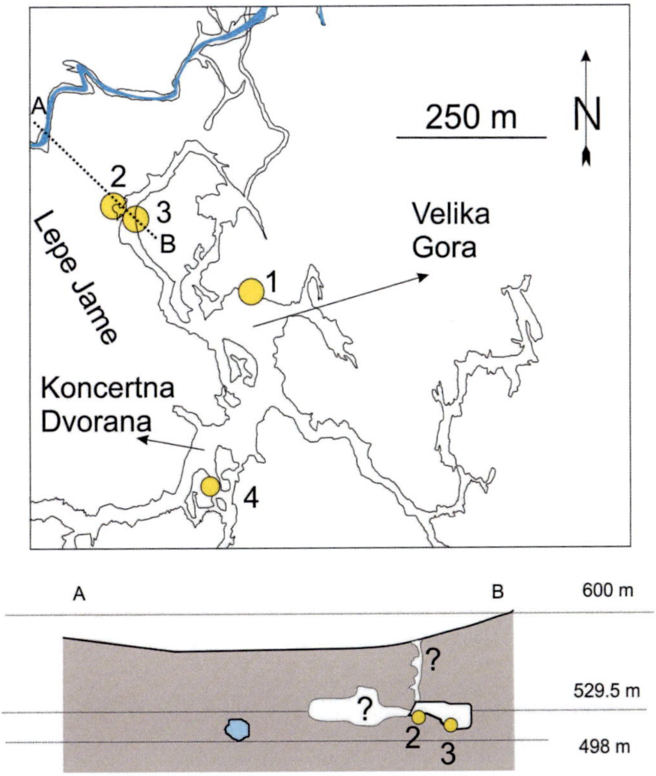

Figure 5.1.3.4.6 Position of Postojna 2 and 3 sites (in yellow circles) and AB cross section with known and unknown (?) passages in horizontal and vertical direction.

into the monitoring site area. Fig. 5.1.3.4.6 presents a cross-section AB and position of known and possible but unknown passages in the vicinity.

At the Postojna 2 location it was determined in February 1963, when there was a winter air circulation ($T_{out} < T_{cave}$), that cave air was flowing toward the surface through an unknown blowing-hole (Gams, 1970). During May 21 to June 7, 1963, at the Postojna 2 location there was no ventilation (Gams, 1970). This is visible also in Fig. 5.1.3.4.7, when in the summer period of 2019 at the Postojna 2 site there were no changes in air temperature because there was no ventilation. In the first months of 2019, air temperature at the Postojna 2 site was generally about 0.1—0.2°C higher than at the Postojna 3 site (Fig. 5.1.3.4.7), while in the autumn of 2019 it was ~0.5°C lower. During the first months of 2019, the Postojna 3 site had a lower air temperature than the Postojna 2 site because colder outside air penetrated into the Lepe Jame passage causing lowering of air temperature values at the Postojna 3 site. In December 2019, air temperature at the Postojna 3 site was more comparable to the Postojna 2 site because colder air was transmitted from Postojna 3 to Postojna 2 site.

Based on the airflow direction with respect to the upper cave entrance (UE) and lower entrance (LE), three ventilation modes were distinguished by Faimon et al. (2012): upward airflow mode (UAF), downward airflow mode (DAF), and transitional mode. In the case of the Postojna 2 monitoring site, the upward airflow mode (UAF) during winter was most pronounced (Fig. 5.1.3.4.8), while summer conditions with a lack of

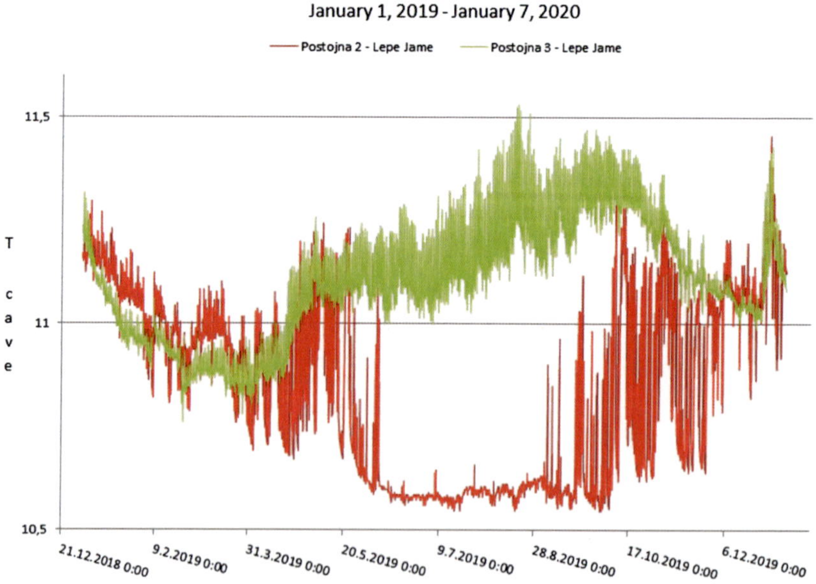

Figure 5.1.3.4.7 Air temperature at Postojna 2 and 3 sites January 1, 2019—January 7, 2020 in °C.

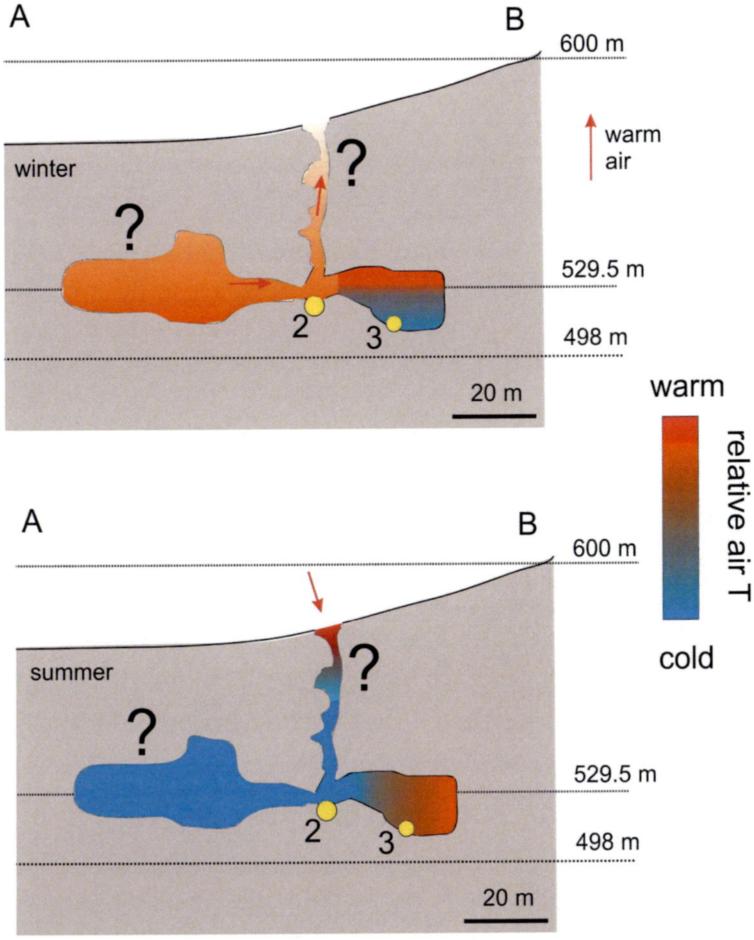

Figure 5.1.3.4.8 Characteristics of winter and summer conditions at Postojna 2 and 3 sites in Lepe Jame passage. ? — unknown passages, 2 — Postojna 2, 3 — Postojna 3.

ventilation resulted in a "cold air-trap" situation, exchanging no air with the surrounding atmosphere. From June to September 2019, the air temperature at the Postojna 2 site increased by +0.02°C, which means that it was stabilized for 3 months at 10.6°C. Functioning of the Postojna 2 site as a cold air-trap during summer can be compared with Črna Jama (see Chapter 5.1.4, Črna and Pivka Jama). Winter ventilation, where relatively warmer cave air (from unknown passages) heats the Postojna 2 site (UAF mode according to Faimon et al., 2012) and exits the cave through a subvertical passage or a system of opened cracks, shows that this site is

Table 5.1.3.4.2 Difference between mean air temperature at Postojna 2 and outside (2010–2019) in °C.

	Mean 2010–2019	Δ*T* (surface – cave)
Postojna 2	10.47	+0.58
Outside	9.89	

different from the Črna Jama site. At the Postojna 2 site there were changes from the UAF mode (Faimon et al., 2012) in winter to the cold air-trap mode in summer, while at Črna Jama there were changes from the downward airflow mode (DAF) in winter to the cold air-trap mode in summer. The air temperature characteristics at the Postojna 2 site show a complex heat exchange situation between the Lepe Jame passage and unknown horizontal and vertical passages behind the Postojna 2 site.

The mean annual air temperature at the Postojna 2 site (see Table 5.1.3.2.1 in Chapter 5.1.3.2, Velika Gora) in 2019 was 10.83°C (it was 11.54°C at Velika Gora and 11.1°C at Postojna 3). The temperature range in 2019 at Postojna 2 was 0.91°C (it was 0.58°C at Velika Gora and 0.77°C at Postojna 3).

The mean air temperature at the Postojna 2 site for the period 2010 to 2019 was + 0.58°C higher than outside the cave (Table 5.1.3.4.2). This value is lower than at Velika Gora (+1.39°C) and the Postojna 3 site (+0.78°C).

In Chapter 5.1.3.2 (Velika Gora), the mean annual air temperature at the Postojna 2 site (Fig. 5.1.3.2.6) was obtained using a manual thermometer (Fig. 5.1.3.4.2) for the period 2004–19, and compared with automatic temperature measurements. The mean air temperature value for 2004–19 determined from manual readings was 10°C for the Postojna 2 site. The rise in the mean annual air temperature from 2004 to 2019 was + 0.62°C. Automatic measurements demonstrated 0.2–0.6°C higher mean annual air temperature values than manual readings at the Postojna 2 site.

Carbon dioxide monitoring

As has already been mentioned in Chapter 5.1.3.3 [Lepe Jame (Postojna 3 site)], periodical studies of carbon dioxide at the Postojna 3 site have been performed during holiday periods and also during periods without high visitor numbers (Šebela et al., 2013; Prelovšek et al., 2018). However, carbon dioxide measurements at the Postojna 2 site have not been published previously. Here, we present two monitoring periods at the Postojna 2 site. The first is a 5-day measuring period (May 28–June 1, 2015) with carbon dioxide monitoring at hourly intervals organized in parallel at the

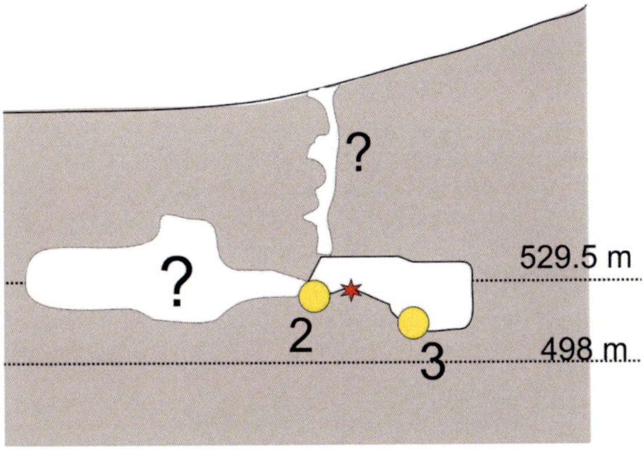

Figure 5.1.3.4.9 Locations of carbon dioxide monitoring in Lepe Jame passage Postojna 2 monitoring site (2), Postojna 3 monitoring site (3) and monitoring site between Postojna 2 and 3 (red star). Unknown passages are marked with sign "?".

Postojna 2 and 3 sites. The second period of carbon dioxide measurement covers October–November 2019 and January–March 2020, when monitoring was organized at the Postojna 2 site and additionally at a new location situated 10 m east of the Postojna 2 site (red star on Fig. 5.1.3.4.9).

Parallel carbon dioxide and air temperature monitoring at the Postojna 2 and 3 sites during a 5-day spring period (Fig. 5.1.3.4.10) show that carbon dioxide concentrations at the Postojna 2 site reached a maximum of 3990 ppm on May 31, 2015, at 6:00 a.m. (Fig. 5.1.3.4.11), while the maximum carbon dioxide concentration at the Postojna 3 site of 2180 ppm was reached from 8:00 a.m. to 20:00 p.m. on every day of the monitoring

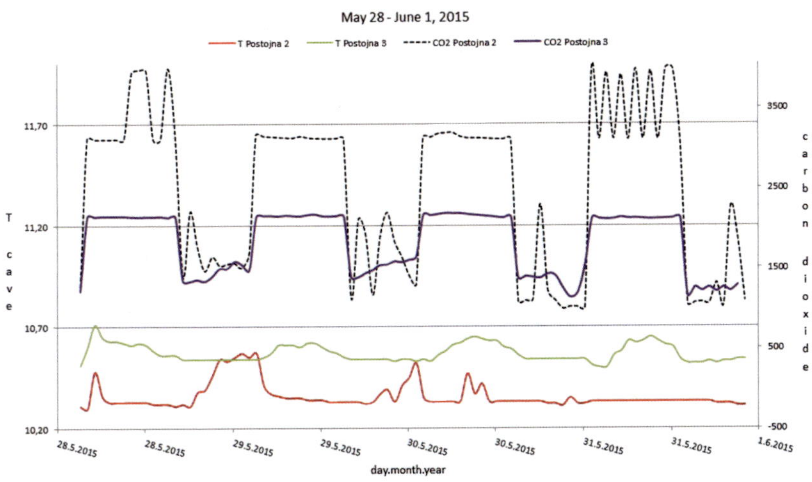

Figure 5.1.3.4.10 Carbon dioxide (ppm) and air temperature (°C) monitoring at Postojna 2 and 3 sites 28 May–June 1, 2015.

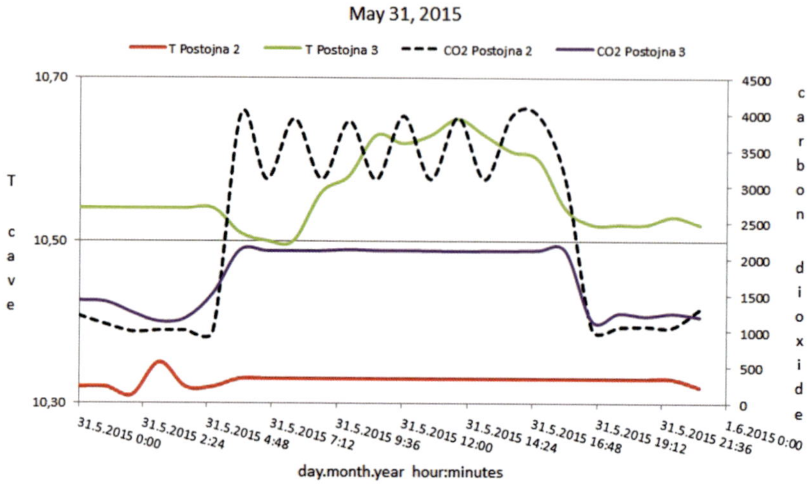

Figure 5.1.3.4.11 Carbon dioxide (ppm) and air temperature (°C) monitoring at Postojna 2 and 3 sites on 31 May, 2015.

period. Daytime hour concentrations were higher than those obtained at night at both locations. The maximum difference in carbon dioxide concentrations at the Postojna 2 site was reached on May 31, 2015 (Fig. 5.1.3.4.11) when at 5:00 a.m. the concentration was 990 ppm and at 6:00 a.m. it had increased to 3990 ppm. In 1 h of the early morning of May 31, 2015, the carbon dioxide concentration increased to 3000 ppm at the Postojna 2 site. On the same day at the Postojna 3 site, the carbon dioxide concentration started to increase at 4:00 a.m. (1180 ppm). It reached its daily maximum at 6:00 a.m. (2119 ppm). The increase in carbon dioxide concentration in 2 h was 939 ppm at the Postojna 3 site.

The air temperature at the Postojna 3 site started to rise on May 31, 2015, at 8:00 a.m., when it was 10.15°C. The daily peak was reached at 14:00 p.m., when the air temperature was 10.65°C (Fig. 5.1.3.4.11). In the evening, at 19:00 p.m., the air temperature was 10.52°C at Postojna 3. The air temperature at the Postojna 2 site was constant at 10.33°C from 5:00 a.m. to 23:00 p.m. This shows that the air temperature at the Postojna 3 site started to increase 4 h after the carbon dioxide level started to increase at the Postojna 3 site. The increase in the air temperature at the Postojna 3 site is to some extent connected with visitors entering the cave as first the spring-summer visits start at 9:00 a.m. summer time (which is 8:00 a.m. sun time—the way how our data are presented in the graphs). The increase in the carbon dioxide level at the Postojna 3 site started at 4:00 a.m. and at the

Postojna 2 site at 5:00 a.m., which is 4 and 3 h before the cave was opened for visitors, respectively. This increase cannot be attributed to the visitors but to natural cave conditions, especially in connection with unknown passages behind the Postojna 2 monitoring site (Fig. 5.1.3.4.9).

During the day hours on May 31, 2015 (Fig. 5.1.3.4.11), another interesting change in carbon dioxide concentrations at the Postojna 2 site occurred from 6:00 a.m. to 17:00 p.m., displayed as a sinusoidal curve, showing a periodic oscillation every 2 h from 6:00 a.m. to 14:00 p.m. and every 3 h from 14:00 p.m. to 17:00 p.m. The difference between the bottom and top of the sinusoidal curve represents ~880 ppm of carbon dioxide concentration. Taking into account that the air temperature at the Postojna 2 site was constant between 6:00 a.m. and 17:00 p.m., the cause of the carbon dioxide oscillations cannot be equated with air temperature at the Postojna 2 site.

If we compare outside the air temperature with the carbon dioxide level at the Postojna 2 site (Fig. 5.1.3.4.12), a good correlation between the high daily air temperature outside the cave and the high carbon dioxide concentrations in Lepe Jame passage is seen. This must be connected with an unknown access between the cave and surface and between unknown passages behind the Postojna 2 site and Lepe Jame passage. Additional support for this is seen in Fig. 5.1.3.4.13, where the air temperature (and air circulation) at the Postojna 2 site is higher during night hours than during day hours. During night hours, the air temperature is constantly at 10.3°C, while during the night it can reach 10.57°C, especially when during nights we have $T_{out} < T_{cave}$ which causes winter ventilation (Fig. 5.1.3.4.8) and rising air temperature and decreasing carbon dioxide levels during nights at

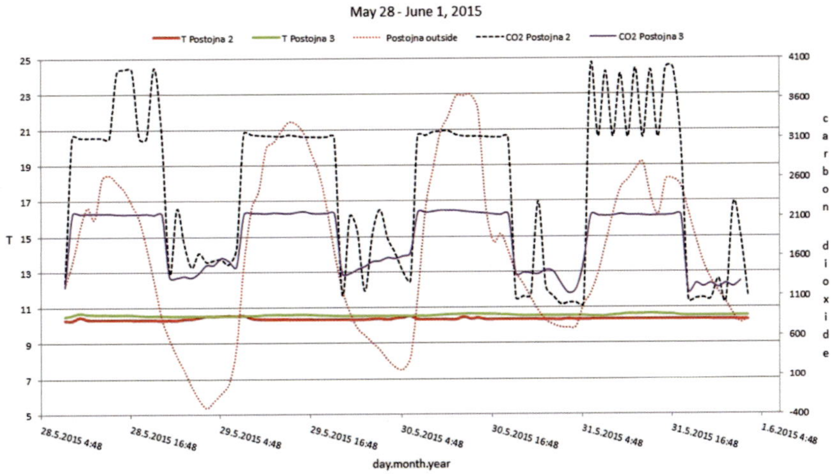

Figure 5.1.3.4.12 Carbon dioxide (ppm) and air temperature (°C) monitoring at Postojna 2, Postojna 3 and outside the cave sites for 28 May–June 1, 2015.

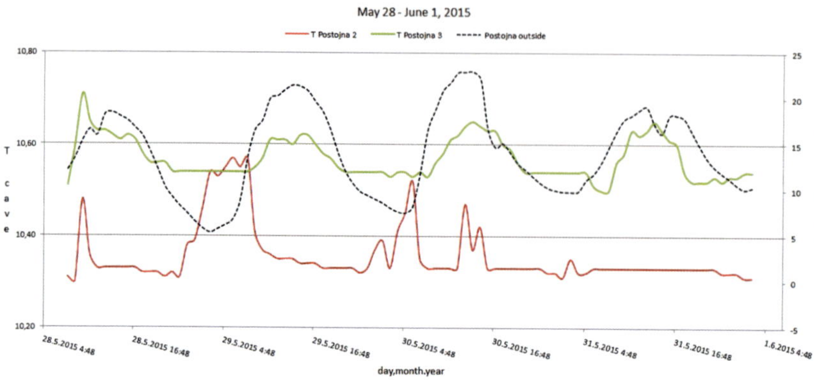

Figure 5.1.3.4.13 Air temperature (°C) monitoring at Postojna 2, Postojna 3 and outside the cave sites for 28 May – June 1, 2015.

the Postojna 2 site. During day hours with a lack of ventilation (summer in Fig. 5.1.3.4.8), carbon dioxide is greatly increased (coming from unknown passages behind the Postojna 2 site), while air temperature is constant, as is typical for the cold air-trap at the Postojna 2 site.

The Pearson correlation coefficient (PCC) between carbon dioxide concentrations at the Postojna 2 and 3 sites for the period May 28, 2015–June 1, 2015 (Table 5.1.3.4.3) gives a very strong correlation (PCC = 0.91) which supports the idea that carbon dioxide concentrations for the calculated period reflect the natural cave conditions at both monitoring sites. An insignificant negative correlation (PCC = −0.17) is calculated for the relation between carbon dioxide concentrations and air temperature at the Postojna 2 site. Some correlation (PCC = 0.68) is shown for the relation between carbon dioxide concentrations at Postojna 2 and the outside air temperature.

The second carbon dioxide data set at the Postojna 2 site represents the period of October 23, 2019 to March 9, 2020 (Fig. 5.1.3.4.14). Air temperature

Table 5.1.3.4.3 Pearson correlation coefficients (PCC) for Postojna 2, Postojna 3 and outside for hourly air temperature and carbon dioxide data May 28, 2015–June 1, 2015.

	PCC
CO_2 Postojna 2—CO_2 Postojna 3	0.91
CO_2 Postojna 2—T Postojna 2	−0.17
CO_2 Postojna 2—T out	0.68
CO_2 Postojna 3—T Postojna 3	0.66
CO_2 Postojna 3—T out	0.73

0.90–0.99 (very strong correlation)
0.70–0.89 (strong correlation)
0.40–0.69 (medium correlation)
0.20–0.39 (weak correlation)
0.19–0.01 (insignificant correlation)
0.00 (no correlation)

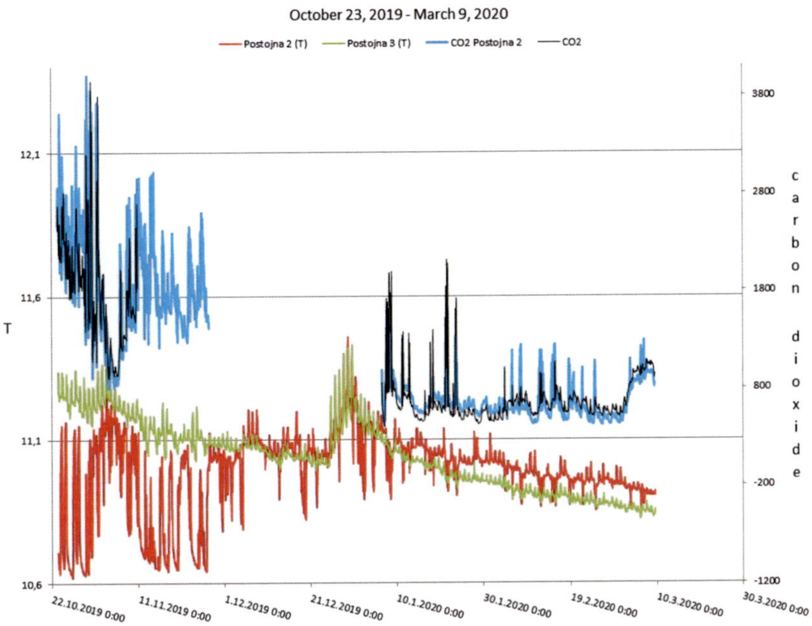

Figure 5.1.3.4.14 Carbon dioxide concentrations (ppm) and air temperature (°C) at Postojna 2, Postojna 3 and "red star (CO_2)" monitoring sites October 23, 2019—March 9, 2020.

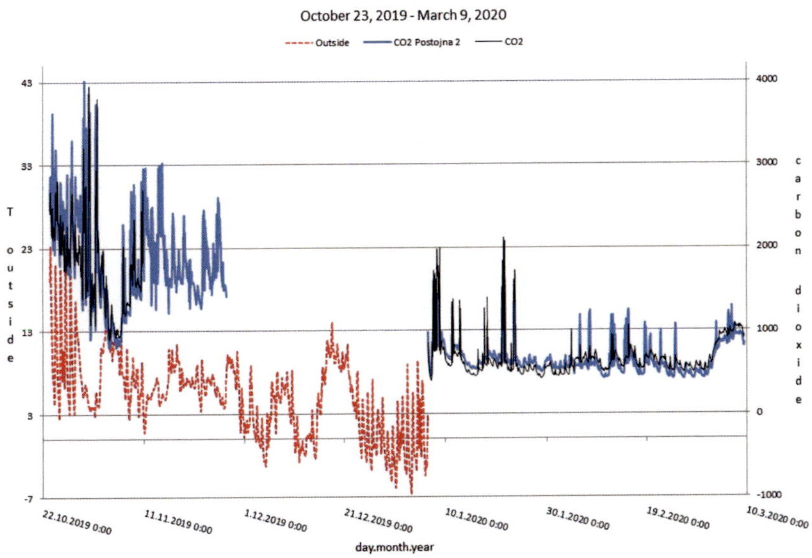

Figure 5.1.3.4.15 Carbon dioxide concentrations (ppm) and air temperature (°C) at Postojna 2, "red star (CO_2)" and outside the cave monitoring sites October 23, 2019—March 9, 2020.

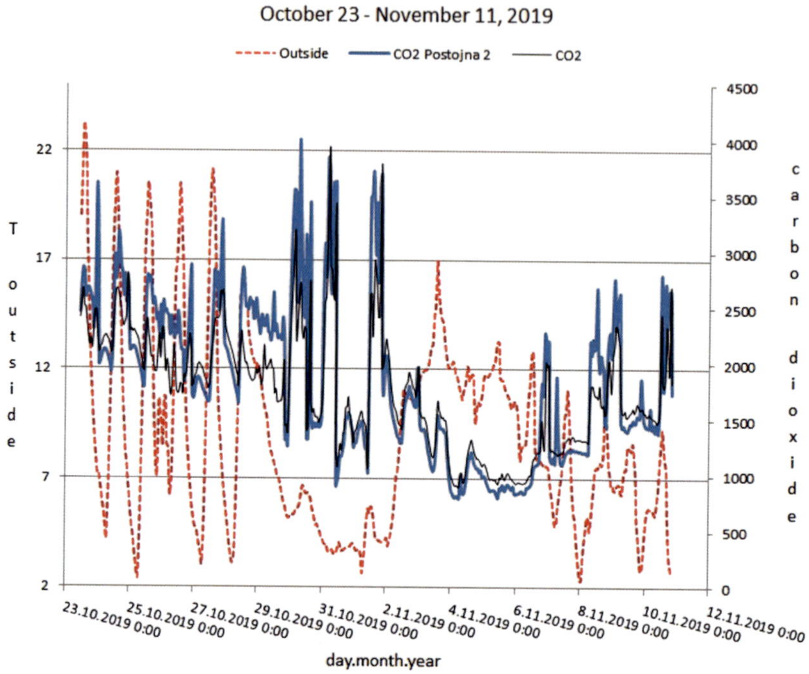

Figure 5.1.3.4.16 Carbon dioxide concentrations (ppm) and air temperature (°C) at Postojna 2, "red star (CO_2)" and outside the cave monitoring sites October 23–November 11, 2019.

is shown for the Postojna 2 and 3 sites, while carbon dioxide is presented for Postojna 2 and the "red star" site near Postojna 2 (CO_2 on Fig. 5.1.3.4.9). It can be seen that both monitoring sites exhibit similar carbon dioxide concentrations during the winter period (Fig. 5.1.3.4.15). In January—March 2020, the background carbon dioxide level is stabilized at ~570 ppm. There are some carbon dioxide peaks, such as on January 21 and 23, 2020, which are higher at the "red star" monitoring site (black curve in Fig. 5.1.3.4.14) than at the Postojna 2 site. On January 21, 2020, at 22:00 p.m., the Postojna 2 site had 710 ppm, while the "red star" site had 2000 ppm. This shows the possibility that carbon dioxide can originate from the Postojna 3 principal Lepe Jame passage and not from the Postojna 2 (Lepe Jame side passage) site.

The October—November 2019 period shows higher carbon dioxide background values in the first months of 2020. In this period, the outside cave air temperature (Figs. 5.1.3.4.15 and 5.1.3.4.16) was much higher than at the beginning of 2020, with daily oscillations of up to 18°C. When the outside air temperature is lower than the cave air temperature ($T_{out} < T_{cave}$), carbon dioxide in the cave increases from 1400 to 4000 ppm (Fig. 5.1.3.4.16). This was the case in the period October 29 to November 2, 2019.

The Postojna 2 monitoring site exhibits very high concentrations of carbon dioxide (up to 4000 ppm), which have a natural origin and are not related to the anthropogenic impact of visitors.

CHAPTER 5.1.3.5

Koncertna Dvorana

Koncertna Dvorana is one of the largest collapsed chambers in Postojna Cave (Fig. 5.1.3.5.1). It has a volume of 50,000 m^3 (Kranjc et al., 2007). The English translation of the chamber's name is Concert Hall, because music concerts have been organized inside it in the past. In 1929, a symphonic orchestra under Pietro Mascagni as the conductor performed the concert inside (Kranjc et al., 2007). This is the last chamber in the cave from which visitors can join the train as it exits. Air temperature monitoring at hourly intervals started here on October 5, 2010, just before the new souvenir shop was constructed in the chamber (Fig. 5.1.3.5.2). An air temperature data logger is placed in about the cave passage that is 5 m wide, 10 m long, and 5−10 m high, and which leads from the collapsed chamber to the tunnel of the underground cave train station (see Fig. 5.1.3.2.1 in Chapter 5.1.3.2, Velika Gora).

Fig. 5.1.3.2.4 in Chapter 5.1.3.2 (Velika Gora) shows the air temperature characteristics in 2019 in Koncertna Dvorana with regard to the outside conditions and other cave sites. The mean annual air temperature in 2019 for Koncertna Dvorana was 11.11°C (see Table 5.1.3.2.1 in Chapter 5.1.3.2, Velika Gora). The temperature range at Koncertna Dvorana monitoring site was within 1.87°C, and was the highest of all the cave monitoring sites in the vicinity. The highest air temperature at Koncertna Dvorana site in 2019 was 12.24°C.

Figure 5.1.3.5.1 Koncertna Dvorana during Christmas—New year period, view towards North. *(Photo: S. Šebela.)*

Natural and Anthropogenic Impacts on Cave Climates
ISBN 978-0-12-822954-5
https://doi.org/10.1016/B978-0-12-822954-5.00014-7

Figure 5.1.3.5.2 Souvenir shop in Koncertna Dvorana. White arrow points to 10 m long cave passage that leads to cave train station below Koncertna Dvorana chamber, view towards SE. *(Photo: S. Šebela.)*

Table 5.1.3.5.1 Difference between mean air temperature in Koncertna Dvorana and outside (2011–2019) in °C.

	Mean 2011–2019	**ΔT cave – surface**
Koncertna Dvorana	10.96	+0.9
Outside	10.06	–

The Pearson correlation coefficient (PCC) for Koncertna Dvorana and outside the cave (see Table 5.1.3.2.2 in Chapter 5.1.3.2, Velika Gora) for 2019 shows a strong correlation (PCC = 0.72). The PCC for Koncertna Dvorana was greater than for the Postojna 1 (PCC = 0.44) and Postojna 3 (PCC = 0.45) sites. The air temperature peak in Koncertna Dvorana occurs 2–4 weeks (between August 12 and 21) after the outside air temperature peak. Spring and summer temperature oscillations in Koncertna Dvorana can reach 0.5°C (Fig. 5.1.3.5.3).

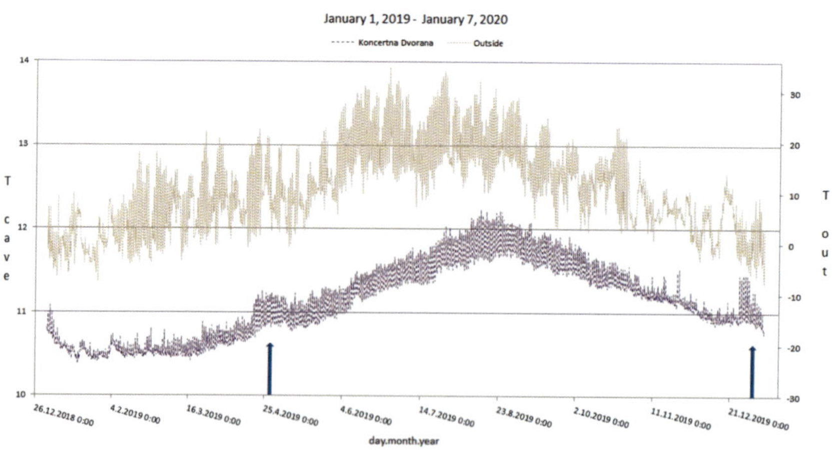

Figure 5.1.3.5.3 Air temperature in Koncertna Dvorana and outside the cave at hourly intervals in °C for January 1, 2019–January 7, 2020.

There are two periods with clear temperature changes at the Koncertna Dvorana site; in Fig. 5.1.3.5.3 these are marked with blue arrows. The first period corresponds to the Easter—May 1st holidays and the second to the Christmas—New Year holidays. Increases in the air temperature of +0.3°C in spring and +0.5°C in winter are a consequence of the increased number of visitors, as already published (Šebela and Turk 2011a; Šebela et al., 2013).

In 2013, the air temperature instrument in Koncertna Dvorana showed a different temperature pattern for the period September 16—November 8, 2013 (Fig. 5.1.3.5.4). The temperature data logger is generally placed in the air ~1.5 m above the cave floor, but during our visit on November 8, 2013, it was placed directly on the cave sediments. The instrument was probably moved onto the floor by one of the visitors, as it was impossible for it to have fallen. Air temperature measurements in spring and summer demonstrated higher daily oscillations (0.3—0.5°C) than after September 16, 2013 (0.1°C). When the air temperature instrument was placed in its regular position it was measuring air temperature, meanwhile, when it was placed on the floor, the measurements were impacted by the ground temperature, which did not show as high daily oscillations as cave air. During our visit to the cave on November 8, 2013, we replaced the temperature data logger in its usual position and kept it running. The instrument detected a temperature drop of 0.23°C. This means that, during November, colder air currents entering the cave from outside lower the air

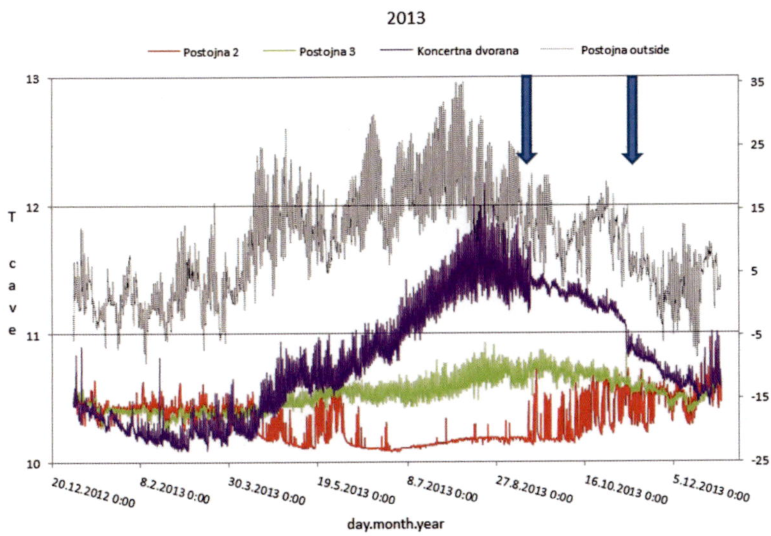

Figure 5.1.3.5.4 Air temperature in Koncertna Dvorana, Postojna 2, Postojna 3 sites and outside the cave in °C for 2013. Blue arrows show the period September 16—November 8, 2013.

temperature in the Koncertna Dvorana site, while the temperature of the cave floor remains at least 0.23°C higher than air at the same location. In November, cave air cools quicker than the natural cave floor in the Koncertna Dvorana site. This monitoring site's position is subjected to a significant impact of colder winter air circulation and also to warmer summer ventilation. It appears that this site is placed within the main cave air circulation currents.

Comparing the air temperature range in Koncertna Dvorana in 2013 (Fig. 5.1.3.5.4) with 2019 (Fig. 5.1.3.5.3), it is clear that year-round air temperature in recent years has been higher than in the past. The increase in mean annual air temperature in Koncertna Dvorana for the period from 2011 to 2019 is + 0.39 degrees, and is lower than at the Postojna 1 (+0.7°C for 2010−19), Postojna 2 (+0.59°C for 2010−19), and Postojna 3 (+0.57°C for 2010−19) sites.

The mean air temperature in Koncertna Dvorana for the 9-year period is 10.96°C (Table 5.1.3.5.1), and therefore is for 0.9°C higher than the outside mean temperature for the same period. This is lower than at Velika Gora (Postojna 1), where it is +1.39°C for the same period, but it is higher than at the Postojna 2 (+0.58°C) and 3 (+0.78°C) sites.

CHAPTER 5.1.3.6

Sepolcro

The air temperature monitoring site Sepolcro is situated at the crossing of Male Jame passage and Stara Jama passage (Fig. 5.1.3.6.1). This Italian name originates from the period before the Second World War which means "tomb" is used because of the speleothem shape (Fig. 5.1.3.6.2). In 1933, the first historic measurements of air temperature started at this location (Crestani and Anelli, 1939). The modern air temperature monitoring site at Sepolcro began operation on March 19, 2015, at hourly intervals. Sepolcro monitoring site was reactivated after 82 years to see the possible differences between air temperature in the past and the current air temperature.

A graph of hourly air temperature measurements at the Sepolcro site over a year is presented together with the outside temperature conditions and those from other cave monitoring sites such as the Postojna 1, 2, 3, and Koncertna Dvorana sites (Fig. 5.1.3.6.3). The Sepolcro site has lower air temperature values than all other cave sites in the vicinity (Velika Gora, Lepe Jame). The mean annual air temperature for 2019 was 10.26°C, with a temperature range of 1.17°C (Table 5.1.3.6.1). Daily temperature oscillations at the Sepolcro site are very low, <0.15°C. In the last days of December 2019 (Fig. 5.1.3.6.3), a very gentle temperature increase of <0.1°C per day was related to a Christmas performance event and increased visitor numbers. Sepolcro has trains with visitors passing daily, with the distance between the railway and the Sepolcro site being about 4 m.

Person correlation coefficients (PCC) for Sepolcro and outside air temperatures for each year between 2016 and 2019 show weak (2016) to medium (2017, 2018, and 2019) correlations (Table 5.1.3.6.2). The cause of this is a time lag between the temperature peak outside the cave and that at the Sepolcro site during the year-long period. The temperature peak at Sepolcro is about 1 month later (August 22, 2019) than the outside (July 25, 2019) temperature peak. If we include the 1-month time lag into the PCC calculation for 2019, we obtain PCC = 0.66 (medium correlation between the Sepolcro air and outside temperature data sets). In Fig. 5.1.3.6.4 we can see that the summer air temperature in 2017, 2018, and 2019 is higher

Natural and Anthropogenic Impacts on Cave Climates
ISBN 978-0-12-822954-5
https://doi.org/10.1016/B978-0-12-822954-5.00015-9

Figure 5.1.3.6.1 Location of Sepolcro monitoring site in Postojna Cave.

Figure 5.1.3.6.2 Sepolcro air temperature monitoring site. White circle shows position of temperature instrument. *(Photo: S. Šebela.)*

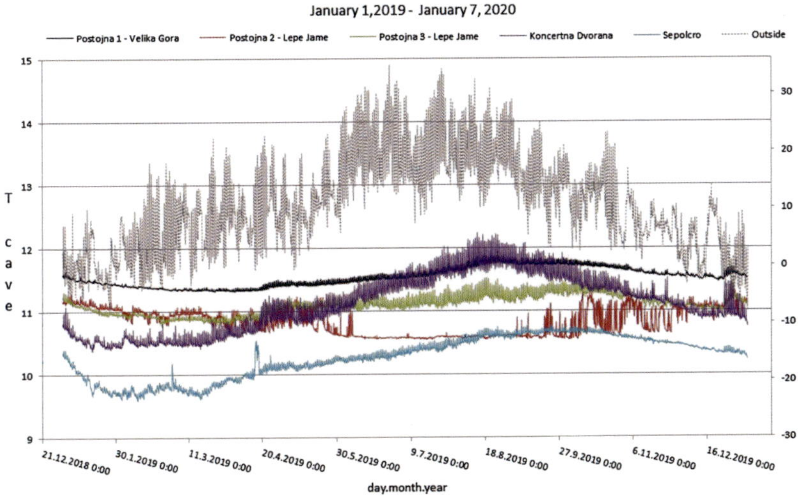

Figure 5.1.3.6.3 Sepolcro air temperature data set in °C for the period January 1, 2019–January 7, 2020 compared with other cave and outside cave monitoring sites.

Table 5.1.3.6.1 Basic statistics for Sepolcro and outside the cave sites for 2019 in °C.

	Sepolcro	Outside
Mean	10.26	10.58
Max	10.76	34.92
Min	9.59	−7.51
Range	1.17	42.43

Table 5.1.3.6.2 Pearson correlation coefficients (PCC) for Sepolcro and outside for hourly air temperature data for years 2016–2019.

	PCC
2016 Sepolcro—out	0.29
2017 Sepolcro—out	0.45
2018 Sepolcro—out	0.45
2019 Sepolcro—out	0.46

0.90–0.99 (very strong correlation)
0.70–0.89 (strong correlation)
0.40–0.69 (medium correlation)
0.20–0.39 (weak correlation)

(<0.2°C) than in 2015 and 2016. Also, the outside air temperature in 2017, 2018, and 2019 was higher than in 2015 and 2016. The higher summer outside air temperatures might have a greater impact on the summer air temperature values at the Sepolcro site, which is seen also in the Pearson correlation coefficients (Table 5.1.3.6.2).

The position of the Sepolcro site at the crossing between the Male Jame and Stara Jama passages is an additional reason why this monitoring site is subjected to air currents from both passages. During the winter, when $T_{out} < T_{cave}$, colder air enters the cave through the bottom of the main cave entrance, following Stara Jama to the inner cave passages. During summer, cave air at Sepolcro is heated not only by the Stara Jama passage, which is the main air conduit, but also the Lepe Jame passage.

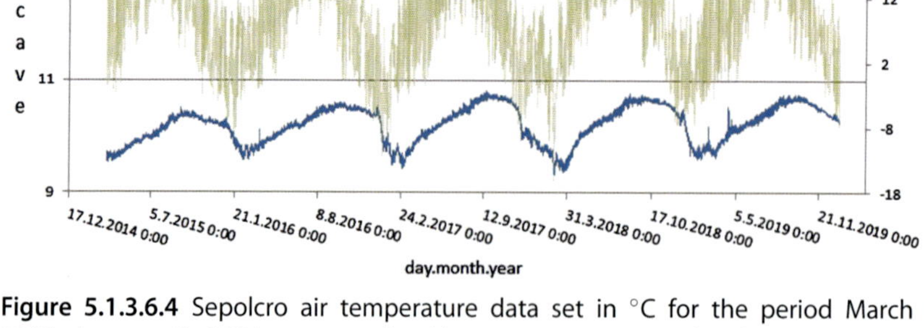

Figure 5.1.3.6.4 Sepolcro air temperature data set in °C for the period March 19, 2015–January 8, 2020 compared with temperature outside the cave at hourly intervals.

Table 5.1.3.6.3 Difference between mean air temperature at Sepolcro and outside (2016–2019) in °C.

	Mean 2016–2019	**Δ*T* (surface – cave)**
Sepolcro	10.22	−0.05
Outside	10.27	–

A statistical calculation of the mean air temperature at Sepolcro and outside, for the period 2016–19, shows that both values are balanced (Table 5.1.3.6.3) and that the Sepolcro site does not overheat comparing with the outside climate.

In the work of Pipan et al. (2019), it was described that Sepolcro (or Razpotje) data, from 19 March 2015 to 4 January 2017, had the highest temperatures in mid-September (10.6°C) and the lowest in January to March (9.6°C). In winter, cold outside air penetrated deep into the cave system and cooled the air at the Sepolcro site, resulting in an overall seasonal pattern similar to that at the surface, albeit with a greatly reduced amplitude (Pipan et al., 2019). Historical cave air temperature data for 1933–37 (Crestani and Anelli, 1939) at Sepolcro showed a temperature range of 7.5–8.4°C. This was 1.74–2.64°C lower than the measurements taken from 2015 to 2017, with an average of 10.14°C (Pipan et al., 2019). What is the reason for the increase in air temperature at the Sepolcro site since 1937? The main cause is related to natural drivers because outside cave temperatures are increasing. However, we cannot exclude the heat brought into the cave by visitors and lights. What the portions are of the natural and anthropogenic causes of the temperature increase at the Sepolcro site remains difficult to determine in such a large system as Postojna Cave.

CHAPTER 5.1.4

Črna and Pivka Jama

Contents

Črna and Pivka Jama are part of the Postojna Cave System (Fig. 5.1.4.1), with their own entrances to the surface. Črna Jama entrance is an opening at the bottom of a collapsed doline (Fig. 5.1.4.2) and faces north to northeast (Fig. 5.1.4.1). Pivka Jama entrance is a shaft about 70 m deep, with about 300 steps that permit tourists to enter the cave (Fig. 5.1.4.3). Črna Jama (Black Cave) become famous in 1797 when Löwengreif found a cave salamander (*Proteus anguinus anguinus*) there. This was probably the place of Valvasor's (1689) descriptions of strange flowstone shapes in the forms of animals and people (Shaw and Čuk, 2015). Within the Postojna Cave system, Črna Jama has the easiest access for visitors to see proteus or "human fish" (*Proteus anguinus anguinus*) in natural cave waters. The cave's name is related to dark coatings deposited over flowstone (Fig. 5.1.4.4). The origin of thin (<1 mm) black deposits in Črna Jama is related to prehistoric forest fires (Šebela et al., 2017) and also to a more recent historic period due to the use of torches in the cave and other forest fires (Zupančič et al., 2011). The youngest growing flowstone is white and can be seen on the metal railing (Fig. 5.1.4.5) where up to 5 cm thick flowstone has accumulated since 1925 when the railing was placed in the cave (Habe, 1986; Šebela et al., 2015a, b).

The natural opening and good access to Črna Jama are the reasons that this part of the cave system was visited (even as organized groups) before the discovery of the more recently discovered parts of the Postojna Cave in 1818.

Natural and Anthropogenic Impacts on Cave Climates
ISBN 978-0-12-822954-5
https://doi.org/10.1016/B978-0-12-822954-5.00016-0

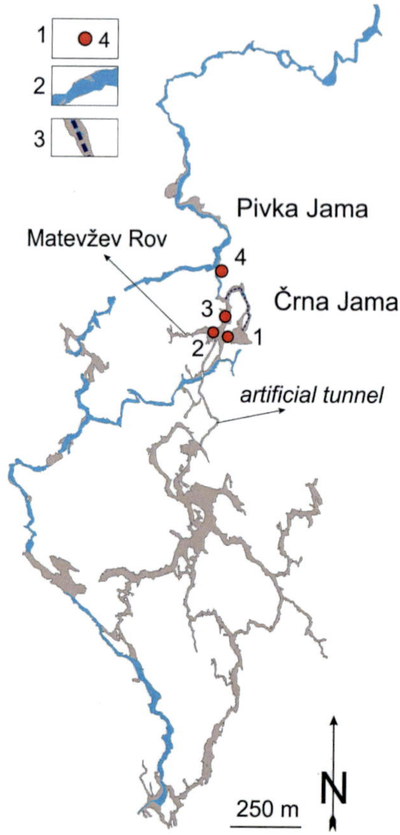

Figure 5.1.4.1 Air temperature monitoring sites in Črna and Pivka Jama. 1-monitoring site (1=Črna Jama 1, 2=Črna Jama 2, 3=Črna Jama 3, 4=Črna Jama 4 or Pivka Jama), 2-underground Pivka water passage, 3-periodic water flow.

On the cave walls are inscriptions from historic periods (Fig. 5.1.4.6). The construction of a 600 m long artificial tunnel from Lepe Jame to Črna Jama (Fig. 5.1.4.1) provided even easier connection between Črna Jama and the rest of Postojna Cave. Some of the old inscriptions (Fig. 5.1.4.7) are not very clear because the walls in Črna Jama can be very moist.

Air temperature monitoring

Air temperature monitoring at Črna Jama started in 2014 in order to compare meteorological conditions in the less visited cave passages (Črna Jama) with more heavily visited parts of the Postojna Cave system. Črna

Figure 5.1.4.2 Entrance to Črna Jama. *(Photo: S. Šebela.)*

Jama has relatively small visitor numbers (3500 visitors per year in the period 2002–2008; Šebela and Turk, 2018) compared to Postojna Cave and is known to be the coldest section of the entire cave system. During warm periods ($T_{cave} < T_{out}$) it acts as a cold air-trap (Fig. 5.1.4.8), exchanging no air with the outside atmosphere and thus preserving the freshness of the cave air even in summer (Šebela and Turk, 2018).

In Črna Jama there are no permanent ice deposits as is the case in some other caves in Slovenia (e.g., Obu et al., 2018). In terms of cave entrance morphology, this cave is similar to Županova Jama (Ravbar and Košutnik, 2014) in central Slovenia with the entrance situated at 440 m, whereas Črna Jama opens at 533 m above sea level. Cold-air traps are commonly of particular ecological interest because they can provide climatic refuges or habitat islands (Edenborn et al., 2012). For instance, in Črna Jama individual bats can be seen hibernating on the cave walls.

Figure 5.1.4.3 Entrance to Pivka Jama. *(Photo: S. Šebela.)*

Figure 5.1.4.4 Black coatings deposited on flowstone in Črna Jama. *(Photo: S. Šebela.)*

Figure 5.1.4.5 Bright flowstone deposited on metal railing in Črna Jama since 1925. *(Photo: S. Šebela.)*

Figure 5.1.4.6 An inscription from 1800 in Črna Jama. *(Photo: S. Šebela.)*

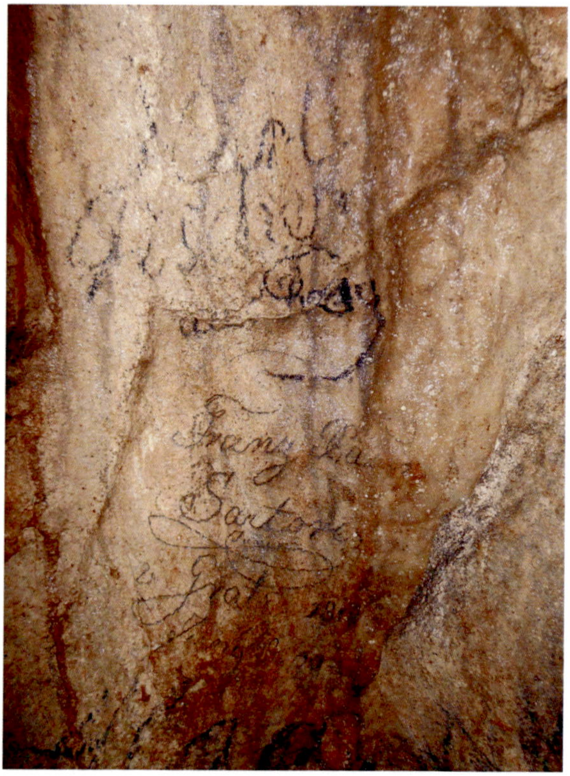

Figure 5.1.4.7 Old inscriptions on the cave wall in Črna Jama. *(Photo: S. Šebela.)*

Črna Jama exhibits winter and summer temperature modes. In winter, dense cold air sinks into the cave, cooling the interior rock surface and cave air. This is due to the accumulation of cold and dry air or unidirectional advection of cold air into the cave during winter (Ford and Williams, 1989; Lütscher and Jeannin, 2004). In winter, because of the density differences between cave and outside air, significant air circulation takes place throughout the cave (Lütscher et al., 2005).

During the warm part of the year, Črna Jama acts as a cold air-trap. Warm outside air is lighter than cold cave air and thus cannot enter the cave, which is at a lower level. There is a lack of significant air exchange with the external atmosphere during the summer season in most caves that are cold air-traps (Lütscher et al., 2005). Warm outside air does not enter through the cave entrance by convection in warm periods, and thus there is a slight increase in the cave air temperature (Črna Jama 2 gradually increases

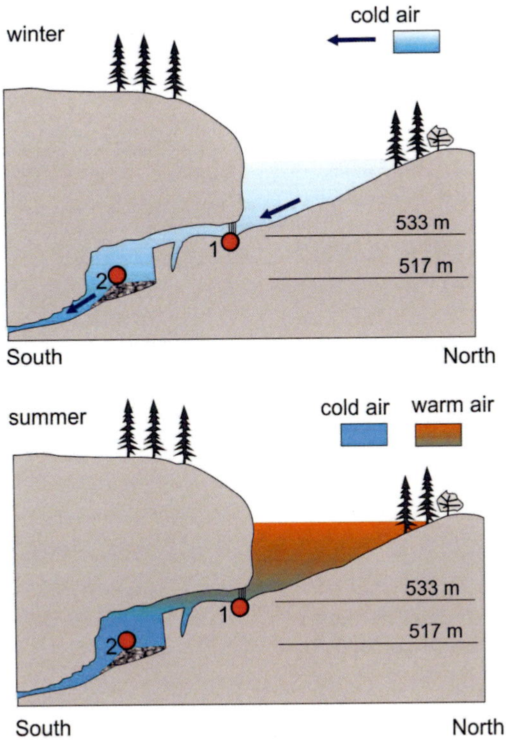

Figure 5.1.4.8 Characteristics of winter and summer conditions in Črna Jama.

from 5 to 7°C) which could be attributed to the heating of the karst massif by conduction (Šebela and Turk, 2018).

The Črna Jama meteorology points to the fact that this part of the Postojna Cave system displays the characteristics of a static cave (Šebela and Turk, 2018). A static cave is characterized as being a pit or subterranean room with no air-drainage potential, and where relatively cold air settles into the cave by gravity during the winter months, whereas the outside ambient air is too warm to sink into the cave during the warm summer months (Williams and McKay, 2015). Static caves should have air currents only at certain times of year. Condensation resulting from warm moist air flowing by cooler cave walls has also been noted (Williams and McKay, 2015). This is also the case in Črna Jama, which is especially visible on cave walls where some old writing has been destroyed by condensation (Fig. 5.1.4.7).

Hourly air temperature measurements have been taken continuously at four locations since January 16, 2014. Three monitoring sites (Črna Jama 1,

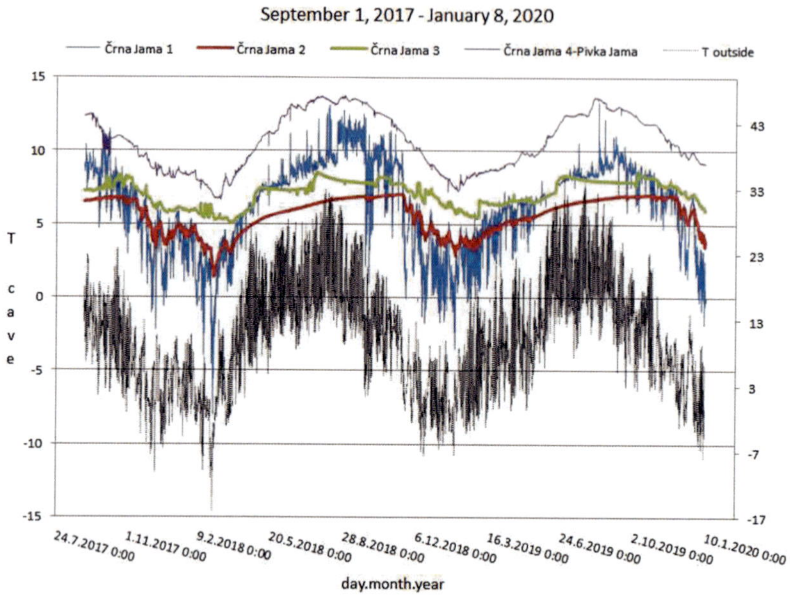

Figure 5.1.4.9 Air temperature datset (°C) at Črna and Pivka Jama compared with outside conditions 2017 to 2020.

2, and 3) are in Črna Jama, and the fourth site (Črna Jama 4) is at Pivka Jama (Fig. 5.1.4.1). The period between 2014 to August 2017 was discussed by Šebela and Turk in 2018. The results from the most recent period of September 2017 to 2020 are shown in Fig. 5.1.4.9.

Črna Jama 1

Črna Jama 1 monitoring site measures temperature conditions at the entrance wire doors (Figs. 5.1.4.1 and 5.1.4.10). At this site, during cold winters, ice stalagmites form. This is caused by water percolating through the 26 m thick limestone ceiling, and entering the cave, where an air temperature below 0°C causes freezing of infiltration water from melting snow above the cave (Šebela and Turk, 2018). The mean annual air temperature at Črna Jama 1 site (Table 5.1.4.1, Fig. 5.1.4.11) was between 5.72°C (2017) and 6.42°C (2018) with a temperature range of between 15.83°C (2016) and 21.11°C (2018).

On November 21, 2018, while visiting Črna Jama, we noticed that the air temperature data logger had been displaced from the original Črna Jama 1

Figure 5.1.4.10 Winter ice stalagmites at Črna Jama 1 site. *(Photo: S. Šebela.)*

monitoring site (Fig. 5.1.4.12) to wire doors (Fig. 5.1.4.13). The instrument had been moved about 4 m east and was in a new location that was more exposed to air currents than at its original site. The instrument was probably moved by visitors. If we compare the air temperature data from August 1 to November 30, 2018, with the same period in 2019 (Fig. 5.1.4.14), we can see that meteorological conditions in 2018 show higher daily oscillations than in 2019. The temporary position of the Črna Jama 1 data logger in 2018 was more exposed to warm air in August and to cold air in November than it was in its original position in 2019, which was hidden behind a brick wall (Fig. 5.1.4.12). The mean annual air temperature at Črna Jama 1 for 2018 was 6.42°C, which is the highest during the entire measurement period (Fig. 5.1.4.11). However, these results are not completely reliable as the data logger was not in the same location for the entire year. Here, we can see how

Table 5.1.4.1 Basic statistics for Črna Jama, Pivka Jama and outside the cave in °C for 2015 to 2019.

	Črna Jama 1	Črna Jama 2	Črna Jama 3	Črna Jama 4-Pivka Jama	Outside
T_{mean} 2015	6.23	5.61	7.06	10.56	10.31
T_{max}	14.06	6.94	8.04	13.4	33.56
T_{min}	−5.7	2.6	5.9	7.1	−10.91
Range	19.76	4.34	2.14	6.3	44.47
T_{mean} 2016	6.16	5.71	7.16	10.36	9.97
T_{max}	13.31	7.06	8.9	13.07	31.31
T_{min}	−2.62	2.43	5.3	6.51	−9.69
Range	15.83	4.63	3.6	6.56	41
T_{mean} 2017	5.72	5.28	6.69	10	10
T_{max}	12.46	6.84	8.35	12.52	37.07
T_{min}	−5.45	1.89	4.55	5.79	−12.31
Range	17.91	4.95	3.8	6.73	49.38
T_{mean} 2018	6.42*	5.48	7.03	10.88	10.55
T_{max}	13.13	7.07	8.52	13.78	33
T_{min}	−7.99	1.38	5.07	6.68	−15.94
Range	21.11*	5.69	3.45	7.1	48.94
T_{mean} 2019	6.17	5.76	7.22	10.61	10.58
T_{max}	13.32	7.05	8.52	13.69	34.92
T_{min}	−3.48	2.86	5.43	7.23	−7.51
Range	16.8	4.19	3.09	6.46	42.43

*Not confidential results due to different position of data logger.

important it is to perform microclimatic monitoring over a long period of time in the same monitoring site. Otherwise, statistic analyses cannot be performed adequately. Unintended movement of the data logger (Črna Jama 1) occurred on August 16, 2018 (18:00 p.m.), and was not remedied until November 21, 2018 (11:00 a.m.).

Črna Jama 2

In summer, the Črna Jama 2 site was used as a teaching example of a cold-air trap within a static cave (Fig. 5.1.4.9). The cave air very slowly warmed before reaching the highest value of 7°C (Table 5.1.4.1) in late October—early November. When the outside temperature dropped below

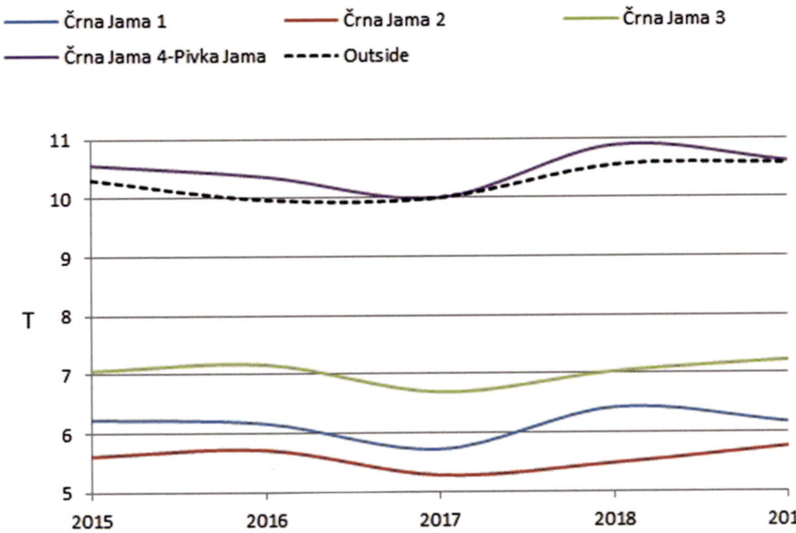

Figure 5.1.4.11 Mean annual air temperature in Črna Jama, Pivka Jama and outside the cave in °C for 2015 to 2019.

Figure 5.1.4.12 Regular position of air temperature data logger Črna Jama 1. *(Photo: S. Šebela.)*

7°C, this caused winter ventilation—the exchange of cold outside air with cave air at the Črna Jama 2 site. The mean annual air temperature at Črna Jama 2 is between 5.28°C (2017) and 5.76°C (2019), and is the lowest in the entire Postojna Cave system.

Figure 5.1.4.13 Position of air temperature data logger Črna Jama 1 in the middle of wire doors in the period August 16 to November 21, 2018. *(Photo: S. Šebela.)*

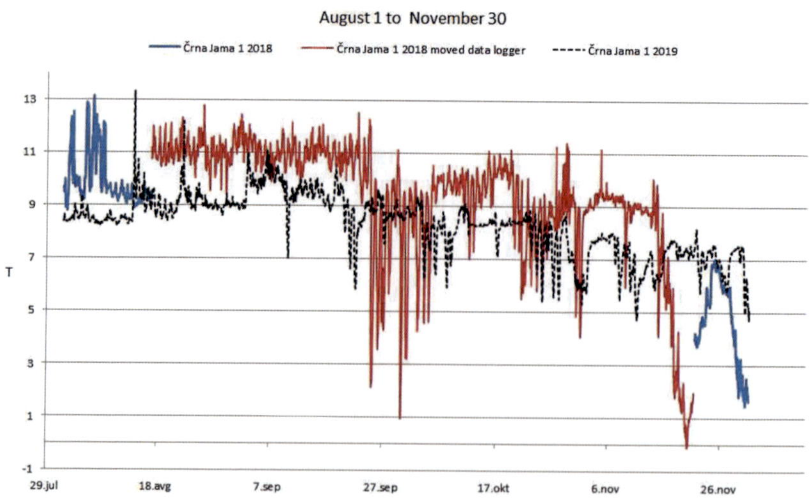

Figure 5.1.4.14 Comparisson between air temperature in °C at Črna Jama 1 site for the period August 1 to November 30 for 2018 and 2019.

The highest temperature peak at the Črna Jama 2 site (October—early November) over a 1-year period lagged outside cave conditions by 3 months (second part of July) (Fig. 5.1.4.15). In the first year (2014) of temperature monitoring, we used Tinytag instruments [Gemini data loggers (UK) Ltd., Orion Group of Companies] with a precision of 0.1°C, which were replaced in 2015 with Baro-Diver (Schlumberger Water Services) data loggers, with an accuracy of ±0.1°C and resolution of 0.01°C. In Fig. 5.1.4.15 this is clearly visible as the Črna Jama 2 site displays a temperature curve with sharp steps in the first half of 2014.

In the period 1933 to 1936, eight air temperature measurements were taken in Črna Jama (Crestani and Anelli, 1939) at a location corresponding to the Črna Jama 2 site. The mean value was 5.27 °C (Crestani and Anelli, 1939). Recent Črna Jama air temperature data (2015—2019, Table 5.1.4.2) indicate a <0.3°C higher temperature than that recorded in historical data since 1933. The Črna Jama 2 site is the most appropriate place within the Postojna Cave system to study long-term climatic changes (Šebela and Turk, 2018) because there is not an excessive number of tourist visits, and human impacts on the cave climate thus are reduced.

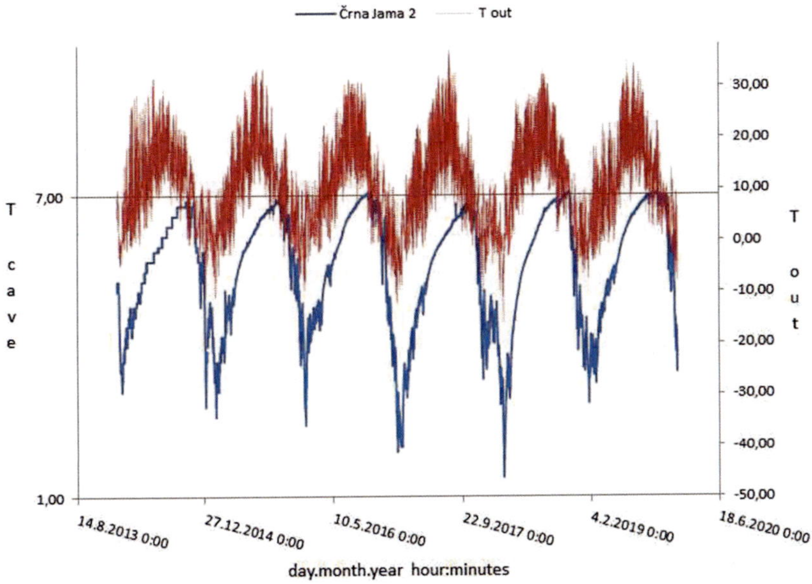

Figure 5.1.4.15 Relation between air temperature in °C at Črna Jama 2 site and outside 2014 to 2019.

Table 5.1.4.2 Difference in °C between mean cave air temperature and mean outside air temperature for period 2015 to 2019.

	Mean 2015–2019	Difference cave – surface
Črna Jama 1	6.14	−4.14
Črna Jama 2	5.56	−4.72
Črna Jama 3	7.03	−3.25
Črna Jama 4–Pivka Jama	10.48	+0.2
Outside	10.28	–

Črna Jama 3

The Črna Jama 3 site (Fig. 5.1.4.9) shows consistently higher (up to about 1.3°C) air temperature values than Črna Jama 2 (Table 5.1.4.2). For the Črna Jama 3 site we generally detect two to three warm events over a 1-year period. This was the case between June 14–16, 2019, when the air temperature increased by almost 1°C and remained at an elevated value. First, it was thought that this is was connected with outside precipitation (Šebela and Turk, 2018) which caused the dry cave passage to be filled with periodic water flow (Fig. 5.1.4.1), meaning that heat from the occasionally flooded branch of underground River Pivka (Fig. 5.1.4.1) might result in warm cave air at the Črna Jama 3 site. A comparison between the hourly air temperature data at Črna Jama 3 with precipitation does not give a clear answer (Fig. 5.1.4.16). Even if it is taken into account that River Pivka needs to flow about 3 km from the ponor (a river sunk into an underground karst system) to reach Črna Jama, there is some time lag in water filling the dry passage after precipitation, and the answer remains unclear for the period June 13–17, 2019 (Fig. 5.1.4.17), but more representative for October 2–9, 2019 (Fig. 5.1.4.18) when, 2 days after the rain, the cave air temperature increased by 0.5°C and 6 days after the rain it had increased by 0.8°C. Additionally, the correlation between Črna Jama 3 and outside air does not give a better response either (Figs. 5.1.4.17 and 5.1.4.18).

Črna Jama 4: Pivka Jama

The Pivka Jama air temperature site (Fig. 5.1.4.1) shows a strong correlation with active water passage of underground River Pivka (Fig. 5.1.4.19). During 2015 to 2019, the mean temperature was very similar to the outside cave conditions (Table 5.1.4.2) and 3.45°C higher than the Črna Jama 3 site and

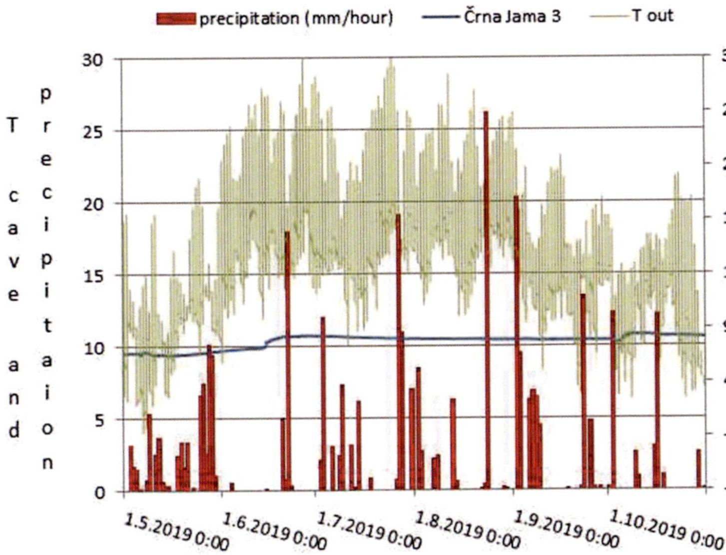

Figure 5.1.4.16 Hourly values of Črna Jama 3 and outside air temperature in °C and precipitations in mm for the period May 1 to October 31, 2019.

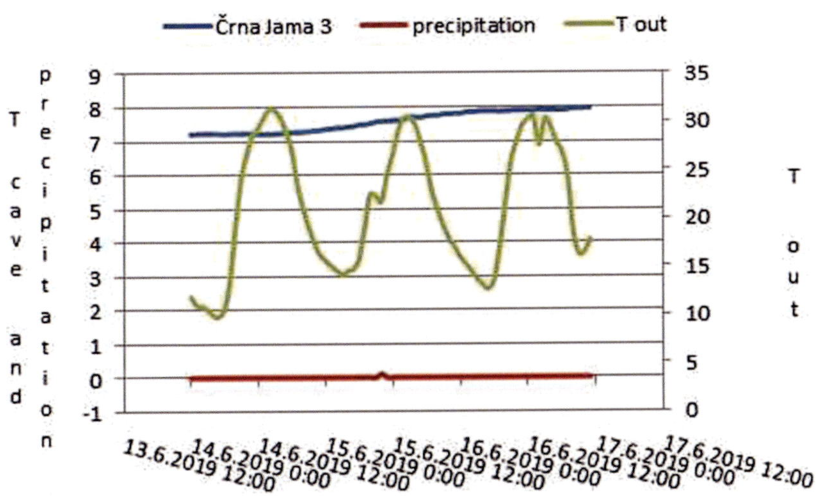

Figure 5.1.4.17 Relations between Črna Jama 3 and outside air temperature in °C and precipitations in mm/hour for the period June 13–17, 2019.

5.12°C higher than the Črna Jama 2 site. The air temperature difference between the cave and outside over a 5-year period (2015–2019) shows that mean values in Črna Jama are 3.25–4.72°C lower than outside (Table 5.1.4.2).

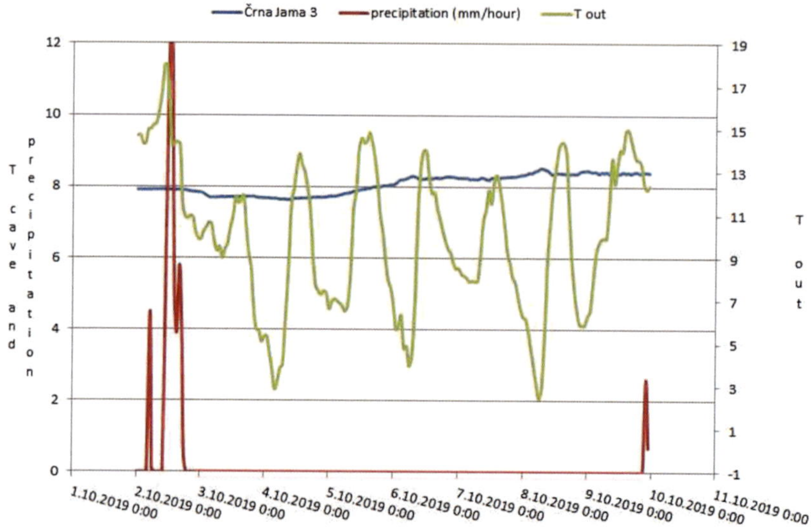

Figure 5.1.4.18 Relations between Črna Jama 3 and outside air temperature in °C and precipitations in mm/hour for the period October 2–9, 2019.

Figure 5.1.4.19 Tourist path and underground River Pivka in Pivka Jama. *(Photo: S. Šebela.)*

As Črna Jama and Pivka Jama are not subjected significant numbers of tourists, the high mean annual air temperature in Pivka Jama (Table 5.1.4.1, Fig. 5.1.4.11) is not due to tourism but to the ventilation between the cave and outside, and also due to the heat transmitted to the cave air by underground River Pivka (Šebela and Turk, 2018).

Correlation coefficients

For the period 2014 to 2017, the air temperature data set was divided according to the different seasons for periodic statistical analyses (Šebela and Turk, 2018). The best correlation was determined for October 16, 2015 to March 17, 2016 for Črna Jama 2–Črna Jama 3, for Črna Jama 2–outside, and for Črna Jama 2–Črna Jama 4 (Šebela and Turk, 2018).

This time total hourly air temperature data for 2019 were used for correlations between cave monitoring sites and outside conditions. The Pearson correlation coefficient (PCC) or the bivariate correlation is a measure of the linear correlation between two variables X and Y. PCC has a value between +1 and −1, where 1 is a total positive linear correlation, 0 is no linear correlation, and −1 is a total negative linear correlation.

Based on year-long hourly measurements (Table 5.1.4.3), a very strong correlation was found between Črna Jama 2 and Črna Jama 3 (PCC = 0.92). A strong correlation was determined between Črna Jama 1 and outside (PCC = 0.78) and between Pivka Jama and outside (PCC = 0.70). No strong correlation was found between Črna Jama 3 and outside (PCC = 0.60) or between Črna Jama 2 and outside (PCC = 0.59). Pearson correlation coefficients for 2019 showed that the Črna Jama 2 and 3 sites both show characteristics of cold-air-trap caves, showing medium correlation with the outside climate.

Matevžev Rov

Matevžev Rov (Matevž passage) represents the western extension of Črna Jama (Fig. 5.1.4.1). It was discovered by Matevž Vilhar from Otok village near Postojna around the year 1900 (Cave Register archives). Morphologically this is a different passage to Črna Jama, which is composed of collapsed chambers and occasional water passages. Matevžev Rov is a dry, generally horizontal, relatively low passage with collapsed chambers. Until May 1956, only the

Table 5.1.4.3 Pearson correlation coefficients (PCC) for Črna Jama, Pivka Jama, and outside for 2019 hourly air temperature data.

	PCC
Črna Jama 1–out	0.78
Črna Jama 2–out	0.59
Črna Jama 3–out	0.60
Pivka Jama–out	0.70
Črna Jama 2–Črna Jama 3	0.92

0.90–0.99 (very strong correlation)
0.70–0.89 (strong correlation)
0.40–0.69 (medium correlation)

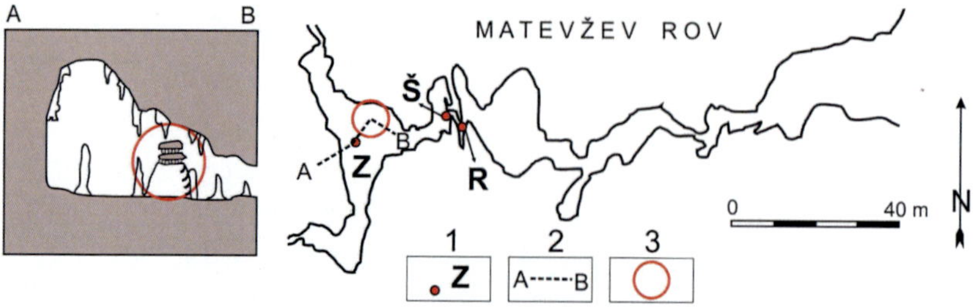

Figure 5.1.4.20 Cave map of Matevžev Rov (Cave Register archives) drawn after 1956 when passage was artificially enlarged near the sign R towards the West with enlarged area near sign Z with flowstone pile. *(1-geodetic points, 2-cross section AB, 3-pile of stalagmites and stalactites.)*

eastern part of the passage was known. On May 16, 1956, Ivan Michler, Franc Hribar, Silvo Modrijan, and Zmago Žele at point R (Fig. 5.1.4.20) artificially enlarged the continuation of the Matevžev Rov toward the west. They reached a smaller room decorated with stalagmites. On the western wall of the room marked Š (Fig. 5.1.4.20) they noticed the inscription: *Matevž Černač from Veliki Otok and Franc Dolenc were here in 1852.* It was partly written in Gothic letters. However this was not the end of the passage, it continued further to the west into a high room decorated with stalagmites, stalactites, and colorful flowstone (the room marked Z in Fig. 5.1.4.20). In the northern edge of the room they found a pile (1 m high and 0.75 m long and wide) of broken or sawn off stalagmites and stalactites (10–15 cm in diameter). This was proof that the stalagmites and stalactites had been being sold without permission (Cave Register archives). People had previously brought stalactites from the eastern part of Matevžev Rov, which was not connected to Črna Jama at that time, to sell them to visitors at the entrance of Postojna Cave (Habe, 1986).

This historic description shows that before artificial extension of Matevžev Rov local people had accessed the western part of Matevžev Rov (from at least 1852 to May 16, 1956) through an unknown vertical passage (30–40 m vertical distance) from the surface. This vertical connection is now blocked. Within the Postojna Cave system there are more such chimney passages that have been opened in historic and/or prehistoric times but that are closed today (Šebela, 2010; Šebela et al., 2017). This is significant for understanding that cave climate conditions have been different in historic and/or prehistoric times than they are today due to changes in air ventilation between the cave and surface caused by natural or artificial alterations to cave entrances.

CHAPTER 5.1.5

Otoška Jama

Contents

Otoška Jama is about 600 m long, mostly horizontal, and is part of the Postojna Cave system with the underground River Pivka at its southeastern edge (Fig. 5.1.5.1). The cave was named after Veliki Otok (Big Island), a village in the vicinity. According to Crestani and Anelli (1939) the cave was discovered on August 18, 1884. Local people noticed that in winter in the area near the cave entrance snow melted more quickly than in other areas and so they suspected that warmer air than the outside air was coming from an unknown cave (Crestani and Anelli, 1939). In 1889 F. Dolenec and J. Vilhar uncovered the entrance by digging (Kranjc and Malečkar, 1988). Today, the entrance situated 531 m above sea level (Fig. 5.1.5.2) is closed by metal doors and at the upper top there is an open space for bats that use the cave for hibernation. The cave allows a limited number of visitors with only individual visits and does not have electric illumination. Archeological remains and animal bones in Otoška Jama have been dated to the Palaeolithic period (Brodar, 1951).

Air temperature monitoring

Air temperature monitoring started on December 4, 2017 (Fig. 5.1.5.3) at two monitoring sites (Fig. 5.1.5.1). While the most visited parts of Postojna Cave have been subject to regular microclimatic monitoring since 2008 and 2009, air temperature monitoring in less heavily visited parts of the cave started not long ago. The first aim in expanding monitoring sites was to provide a new data set from areas with very few tourists and to make comparisons these areas and those cave parts heavily visited. The second aim was to become familiar with microclimatic conditions in non-tourist parts before they became used by tourists. Otoška Jama is mentioned as a possible future speleotherapy site. Speleotherapy uses certain conditions specific to caves and salt-mines to treat several medical conditions, especially respiratory and skin-related disorders (Lăzărescu et al., 2014). If in the future

Natural and Anthropogenic Impacts on Cave Climates
ISBN 978-0-12-822954-5
https://doi.org/10.1016/B978-0-12-822954-5.00017-2

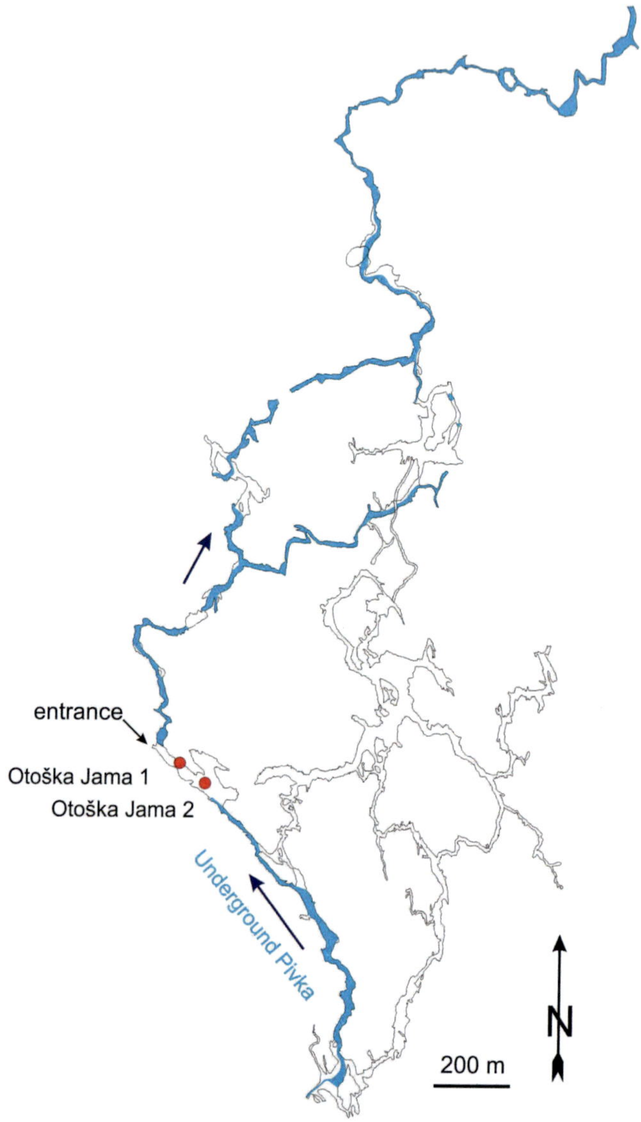

Figure 5.1.5.1 Position of Otoška Jama air temperature monitoring sites within Postojna Cave system.

speleotherapy becomes active in Otoška Jama this will greatly increase the human presence in the cave.

Two monitoring locations show different air temperature ranges on a yearly basis (Fig. 5.1.5.3). The temperature at the Otoška Jama 1 site is lower than at the Otoška Jama 2 site. At the Otoška Jama 1 site there are

Figure 5.1.5.2 Entrance to Otoška Jama. *(Photo: S. Šebela.)*

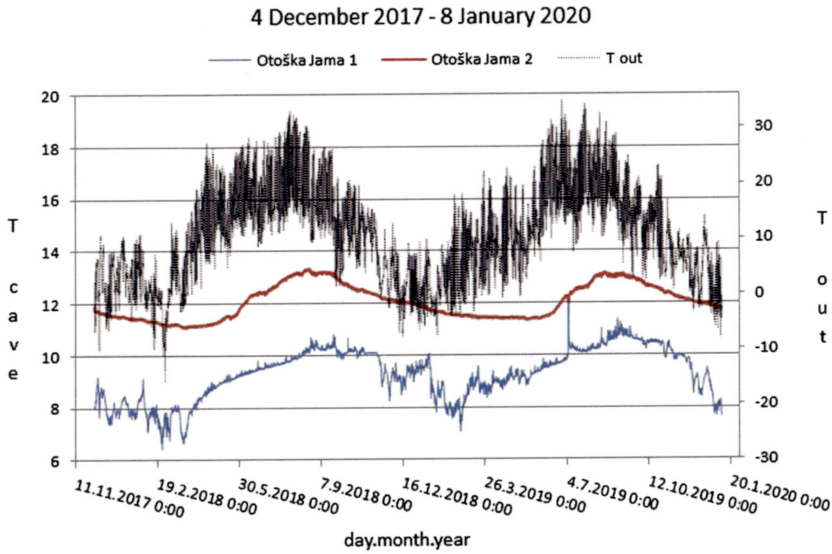

Figure 5.1.5.3 Hourly data-set of air temperature (°C) measurements in Otoška Jama, 4 December 2017 to 8 January 2020.

more pronounced daily air temperature oscillations than at the Otoška Jama 2 site. Also, the difference in mean annual air temperatures between the two monitoring sites shows that the Otoška Jama 1 site had a 2.98°C lower value than the Otoška Jama 2 site in 2018 (Table 5.1.5.1). The difference was even greater in 2019 when it was 3.41°C.

Table 5.1.5.1 Basic statistics of air temperature in °C for Otoška Jama.

	Otoška Jama 1	Otoška Jama 2	Outside
2018			
T mean	9.1	12.08	10.55
T max	10.82	13.3	33
T min	6.42	11.1	−15.94
Range	4.4	2.2	48.94
2019			
T mean	9.54	12.95	10.58
T max	12.21	13.15	34.92
T min	7.07	11.35	−7.51
Range	5.14	1.8	42.43

Low winter air temperature values at Otoška Jama 1 are due to the closeness of the monitoring site to the cave entrance. Dense and heavy cold winter outside air reaches the Otoška Jama 1 site and causes a drop in cave air temperature (~2°C) during winter. The difference between the highest summer values and lowest winter values is even higher (4.4°C in 2018 and 5.14°C in 2019). The highest summer values at Otoška Jama 1 in 2018 were 10.82°C, and 12.21°C in 2019. This last value of 12.21°C, which was reached on July 5, 2019, at 6:00 a.m. is particular (Fig. 5.1.5.4). The sudden increase in air temperature started on July 4, 2019, at 11 a.m. and finished on July 5, 2019, at 10 a.m. During the same period, the Otoška Jama 2

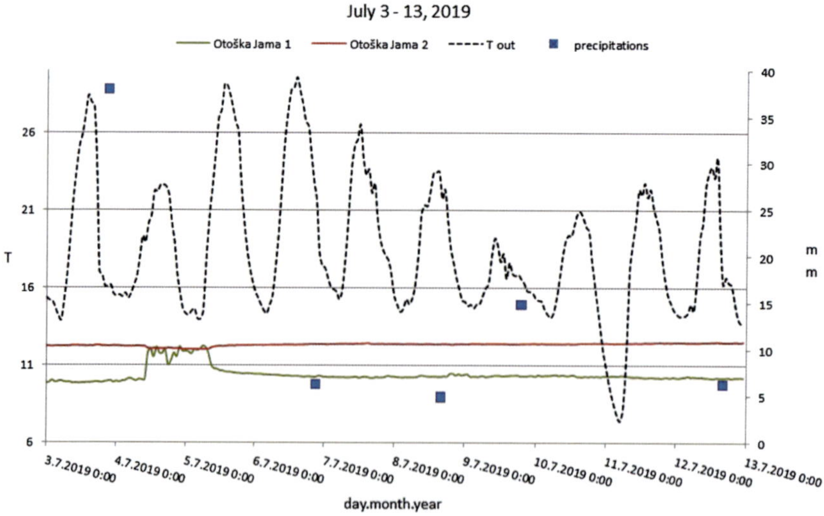

Figure 5.1.5.4 Air temperatue (°C) in Otoška Jama and outside with precipitations (in mm per day) for the period July 3–13, 2019.

monitoring site described a small decrease in air temperature. The outside air temperature is not a sufficient explanation for this effect, therefore we need to exclude errors in measuring instrument, because it is unlikely that both air temperature devices had errors at the same time.

On July 3, 2019, between 16:30–19:00 p.m. there was 38 mm of precipitation in Postojna. The next day, July 4, 2019, the outside air temperature (Fig. 5.1.5.4) reached a maximum value of 22.6°C, which was 5.8°C lower than the day before. The reason for this was a rain event in the afternoon of July 3, 2019. In the following days (Fig. 5.1.5.4) there were some minor precipitation events and a greater drop in air temperature on July 11, 2019, with a minimum value at 5:00 a.m. However, the event observed on July 4–5, 2019, was not repeated. It appears that rain precipitation of 38 mm on July 3, 2019, caused the temporary increase (July 4, 2019 at 11 a.m. to July 5, 2019 at 10 a.m.) in air temperature at the Otoška Jama 1 site by ~2°C after 35 h.

The Otoška Jama 2 site had minimum winter air temperature values of 11.1°C in 2018 and 11.35°C in 2019. This shows that there is not a significant impact of outside winter air on the Otoška Jama 2 site. It must not be forgotten that this site is situated close to the underground River Pivka passage. However, if we take a look at Pivka Jama (Chapter 5.1.4), where the Črna Jama 4 monitoring site is also close to the underground River Pivka, we notice that the mean annual air temperature in 2019 was 10.61°C and in 2018 it was 10.88°C, which is lower than at Otoška Jama 2, where in 2019 the mean annual air temperature was 12.95°C and 12.08°C in 2018. The difference between the monitoring sites in mean annual air temperature is related to the underground River Pivka path and the heat transmitted into cave air from the river. The Otoška Jama 2 site is about 1 km from Pivka ponor—the entrance point of the surface River Pivka into the Postojna Cave system—while the Črna Jama 4 monitoring site is about 3 km from the Pivka ponor. This means that the air temperature of the underground River Pivka cools down on a 3 km long cave path and has less of an impact on the cave air and rock in Pivka Jama than it has in Otoška Jama. This effect can be explained by thermal retardation in karst aquifers (Luhmann et al., 2015; Kaufmann et al., 2016) and describes water movement through karst aquifer and heat exchange between water and rock. Another possibility is that the position of the Otoška Jama 2 site can have characteristics of a warm air trap which preserves the high air temperature not only in summer but also in winter (Fig. 5.1.5.3).

Table 5.1.5.2 Mean air temperature (°C) for the period 2018–2019 in Otoška jama and outside and difference between cave and surface values.

	Otoška Jama 1	Otoška Jama 2	Outside
2018–2019	9.32	12.51	10.56
ΔT (cave − surface)	−1.24	+1.95	–

Regarding the mean air temperature for the period 2018–19 (Table 5.1.5.2), we become aware that Otoška Jama 2 is the warmest of all the monitoring sites in this entire cave system when compared with the mean outside air temperature. At the Otoška Jama 2 site the air temperature is +1.95°C higher than outside. It must not be forgotten that Otoška Jama receives very low visitor numbers and can be treated as almost a non-touristic cave. Here we deal with natural microclimate conditions and especially with the fact that a lot of heat is transmitted to the cave air from the underground River Pivka and that the monitoring site probably represents a warm air trap.

CHAPTER 5.1.6

Visitors to Postojna Cave

The first visitors to Postojna Cave date back to prehistoric times (Šebela, 2019). These first visitors would probably be better described as but cave dwellers. The cultural layers from Postojna Cave belong to the last half of the Middle Paleolithic (roughly from 160,000 to 40,000 BP), which is the end of the Mousterian (Brodar, 1966), when Neanderthals used it as a shelter, especially during relatively cold and humid periods (Turk, 2011). At the beginning of the Würm Ice Age (115,000 to 11,700 years ago) the presence of Paleolithic man in Postojna Cave can be proved reliably (Brodar, 1966). Cultural layers in Postojna Cave have been found in Otoška Jama, Čarobni Vrt, Rov pri Slonovi Glavi, and Postojna Cave main entrance areas (Fig. 5.1.6.1). Some old entrance passages are today blocked with fluvial sediments, but in the past they were opened and used by prehistoric humans. The best known prehistoric site in the vicinity of Postojna Cave is Betalov Spodmol Cave (Bavdek and Josipović, 2017), which is one of the best known Paleolithic cave locations in Slovenia. Pre-Mousterian and Mousterian cultural layers have been found in Betalov Spodmol, in addition to the remains from Würm Ice Age and the Neolithic (Brodar, 1966; Bavdek and Josipović, 2017). he uppermost layer contained ceramic and animal bone remains from the Bronze and Iron Ages (Brodar, 1966; Bavdek and Josipović, 2017). This shows that the area around Postojna Cave has an extremely rich history of prehuman and human settlements.

Inscriptions in Rov Starih Podpisov (Fig. 5.1.6.1) date back to 13th century (Gams, 2003; Čuk, 2008; Kranjc et al., 2007), with the most famous dated to 1213 (Agapito, 1823), which can no longer be found. In 2013, Postojna Cave celebrated the 800th anniversary (1213–2013) of visits to the cave. On that occasion a numismatic 2 Euro coin with the Postojna Cave silhouette was issued (Fig. 5.1.6.2). It needs to be stressed that before 1818 only a small part of the Postojna Cave system was known (B in Fig. 5.1.6.1), however the cave has already merited its celebrity status.

Over the last six centuries Postojna Cave has attracted scientists, especially karstologists, biologists, and speleologists, and other amateur visitors. Included among these is Valvasor (1689) who visited Črna Jama, and

Natural and Anthropogenic Impacts on Cave Climates
ISBN 978-0-12-822954-5
https://doi.org/10.1016/B978-0-12-822954-5.00018-4

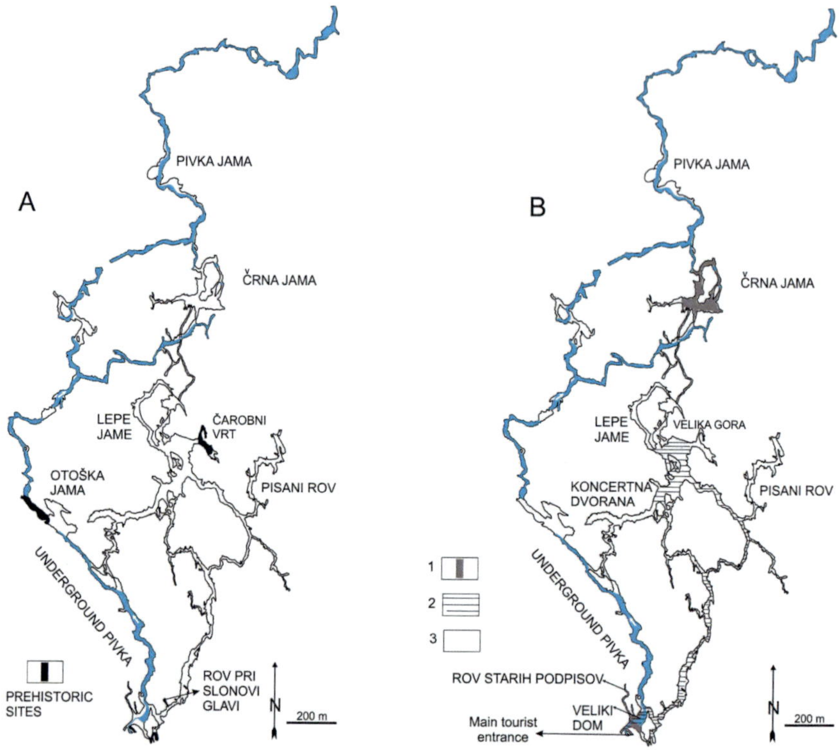

Figure 5.1.6.1 Prehistoric (A) and historic (B) discoveries of the Postojna Cave. Legend for historic (B) discoveries: 1- before 1818, 2- 1818–1819, 3- after 1819.

described strange fantasy forms of speleothems and guided organized tours for his scientific colleagues (Kranjc et al., 2007). At that time Črna Jama was a place where visitors were able to see the cave animal *Proteus anguinus*

Figure 5.1.6.2 Numismatik coin for 800th anniversary (2013) of visits to the Postojna Cave.

anguinus in its natural surroundings (Šebela et al., 2015a, b). In 1748, Nagel draw the first cave map of the Postojna Cave (Kranjc et al., 2007).

The most important year for modern tourism in Postojna Cave was 1818, when local cave lamplighter and guide Luka Čeč discovered extensive new parts of the cave (Šebela et al., 2015a, b). On the occasion of the visit of the Austrian emperor he climbed across the underground River Pivka in Veliki Dom chamber to place an inscription in honor of the noble guest and discovered new dry passages (Gams, 2003; Kranjc, 2004; Čuk, 2008). On August 17, 1819, the Austrian successor to the throne, Ferdinand, was the first official visitor to the new passages. After his visit, the new passages were opened for tourism and the visitors' book was introduced (Kranjc, 2004; Šebela et al., 2015a, b). In 1819, the official counting of Postojna Cave visitors started (Fig. 5.1.6.3).

The period when new passages in Postojna Cave were discovered is connected with turbulent events in Europe. In 1816, severe summer climate fluctuations (probably connected with a massive eruption of Mount Tambora volcano in April 1815 in Indonesia) caused mean global temperatures to decrease, resulting in food shortages across the Northern Hemisphere (known also as a "year without a summer" or "poverty year"). It was the worst famine of the 19th century in Europe. This was an

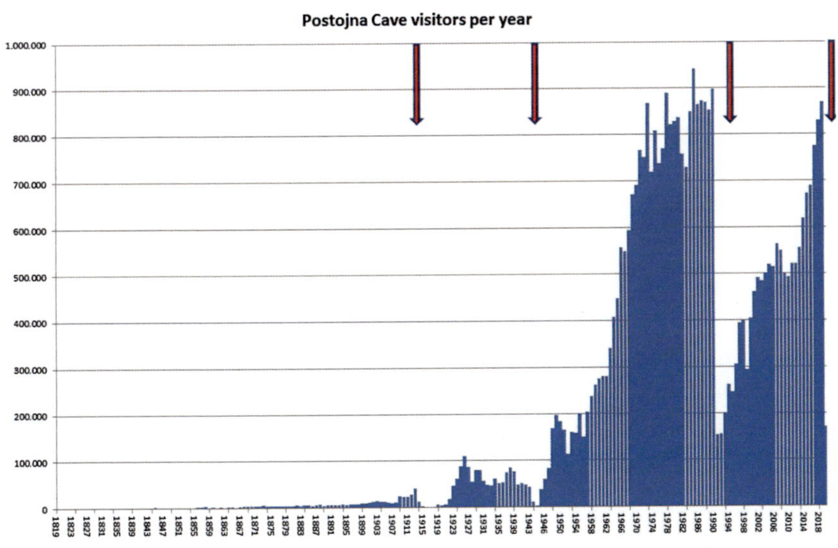

Figure 5.1.6.3 Number of visitors to Postojna Cave (1819–2020). Red arrows mark strong drops in visitor numbers.

important period for local people to develop Postojna Cave as a famous tourist destination (Šebela et al., 2015a, b). The importance of Postojna Cave as a tourist destination was supported with the first guidebook written in Italian (Agapito, 1823).

The number of visitors to Postojna Cave each year (1819–2020) is presented in Fig. 5.1.6.3. There are four large drops in the number of visitors (red arrows in Fig. 5.1.6.3). The first drop is related to the First World War and the second to the Second World War (Šebela et al., 2015a). The third drop is related to the independence of Slovenia in 1991, and the politically unstable conditions in the surrounding region (especially former Yugoslavia), which prevented tourists from visiting the cave (Šebela et al., 2015a). The final drop in 2020 is related to the COVID-19 Coronavirus pandemic, which caused an ~85% drop in visits to Postojna Cave in 2020 as compared with 2019. Here, we can see that global social, economic, and health issues in the last 200 years are the principal causes of large drops in visitor numbers.

The data set of annual visitor numbers (1819–2017) for Postojna Cave was studied from historical and statistical points of view (Brenčič, 2018). It was found that, until 1960, a general trend can be described with an exponential curve model, with periods after this year illustrated by a double logistic curve (Brenčič, 2018). It was shown by Brenčič (2018) that in addition to monotonic curve trends in the data set, more catastrophes were identified, for which the mean arrival time, duration, and intensity were estimated. For future growth trends it was illustrated that the next drop in the visitors curve can be expected when the annual number of visitors is ~890,000 (Brenčič, 2018). In fact, the year 2020 gave credence to this study.

The highest number of visitors in a single day was in August 8, 1985, when there were 12,025 visitors, and the record year was 1985 when 942,000 tourists visited Postojna Cave (Kranjc et al., 2007).

In 2017 Postojna Cave welcomed its 38 millionth visitor since 1819 (Fig. 5.1.6.4) (https://www.postojnska-jama.eu/en/media-centre/news-and-press-releases/postojna-cave-welcomes-its-38-millionth-visitor/). However, on May 29, 2019, the cave had already celebrated 39 million visitors (https://www.rtvslo.si/tureavanture/novice/39-milijonti-obiskovalec-postojnske-jame-na-ogled-200-let-stara-vpisna-knjiga/489480).

On August 15, 2018, there were 7700 visitors to Postojna Cave. Of these, 27% were from Italy, 8.5% from Germany, 8% from Spain, 6.5% from Israel, and 5.5% each from the Netherlands and France (http://www.

Figure 5.1.6.4 Main tourist entrance to Postojna Cave. In 2017 the cave welcomed 38 million visitors since 1819. *(Photo: S. Šebela.)*

rtvslo.si/tureavanture/novice/park-postojnska-jama-podrl-mejnik-14-000-obiskovalcev-dnevno/463285).

In 2017, 16% of visitors were from Italy, 8% from Germany, 7.5% from Slovenia, and 7% from South Korea (https://www.postojnska-jama.eu/sl/medijsko-sredisce/novice/rekordno-leto-v-postojnski-jami/). Other Asian countries (China with Hong Kong, Taiwan, Japan) also contributed numerous visitors (https://www.rtvslo.si/tureavanture/podobe-slovenije/postojnska-jama-leto-koncuje-z-zivimi-jaslicami/475436). In 2019, 2.4% of visitors were from China, 2.4% from Hong Kong, and 3.7% from Taiwan (https://siol.net/posel-danes/novice/zaradi-smrtonosnega-virusa-kitajci-odpovedujejo-potovanja-v-slovenijo-518560).

The highest monthly number of visitors is in August. There were more than 114,000 visitors in August 2007, representing ~23% of the annual number of visitors. The lowest visitor numbers for the period 2004–11 were in January and February (Fig. 5.1.6.5). After 2011, visits to Postojna Cave increased during colder months also. In recent years the number of visitors almost doubled in the period 2004–11 (Fig. 5.1.6.3). In July 2011 there were 90,000 visitors, and on August 17, 2019, Postojna Cave welcomed 7800 tourists, which is the highest daily number of visitors in recent years.

Daily visitor numbers in July 2020 were up to 1000, what is about 70%–80% less than July 2019, due to the COVID-19 Coronavirus pandemic.

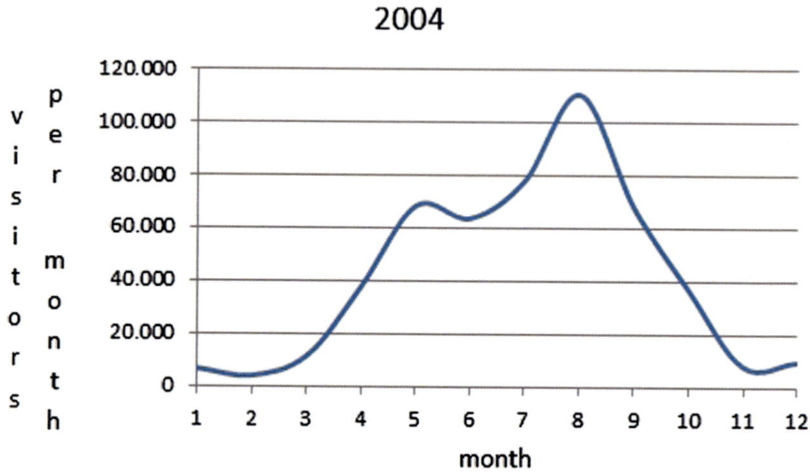

Figure 5.1.6.5 Monthly number of visitors in 2004 in Postojna Cave.

Increased number of visits to Postojna Cave at the end of December are related to Christmas performance events. In December 2019, it was the 30th anniversary of this event during which about 25,000 people visited the cave in 5 days.

Historic visits by monarchs and nobles to the show caves were analyzed by Shaw and Čuk (2002), who described 103 visits to Slovene caves by emperors, kings, and other rulers. The visitors' book for Postojna Cave started in 1819, and includes names of those who visited the cave each year (Shaw and Čuk, 2002). From 1857, imperial and royal visitors were no longer recorded in the ordinary visitors' book but in special a Memorial or Golden Book. On the first page there is a signature of the Emperor Franz Joseph of Austria and below it a signature of Empress Elisabeth (the famous "Sisi") dated March 11, 1857 (Shaw and Čuk, 2002).

After some preliminary attempts in 1863, electric lighting was first used in Postojna Cave in 1883 for a visit by Emperor Franz Joseph. The permanent electric installation of 1884 in Postojna Cave was the third anywhere in the world (Shaw, 2003).

CHAPTER 5.1.7

Anthropogenic and natural impacts on Postojna Cave microclimate

Having been a show cave for more than 200 years, the Postojna Cave microclimate is subjected to anthropogenic and natural impacts. It has been shown that during increased visitor numbers in winter (such as during Christmas performance events in the last days of December) the water temperature of the rimstone pool in Lepe Jame passage retains the increased water temperature for a longer period than is the case for air temperature (Šebela and Turk, 2014a). Anthropogenic impacts on cave climate parameters due to visitors are detected in air temperature and also carbon dioxide concentrations. However, these are mostly short-term impacts because after visitors leave the cave, the air temperature and carbon dioxide concentrations return to the previsit values (Prelovšek et al., 2018). Generally, in Postojna Cave we do not find a cumulative increase of air temperature and carbon dioxide. The cave is well ventilated (Šebela and Turk, 2011a), which causes dilution of carbon dioxide and recovery of the cave climate conditions. In smaller cave chambers, such as Biospeleološka Postaja, which has ~130,000 visitors per year (number from 2019), higher air temperature and carbon dioxide values are observed than in the natural state. With cooperation between the cave manager, Ministry of the Environment, and cave guardian (an institution with strong karstology knowledge that suggests guidelines for sustainable use of the cave for tourism), the air temperature values and carbon dioxide concentrations have been reduced at Biospeleološka Postaja chamber by adopting a forced ventilation (door opening) solution.

In many cases it is not easy to distinguish the ratio between anthropogenic and natural impacts on cave climate. We know that the highest air temperature peak in the most visited passages of Postojna Cave is in the second half of August (Šebela and Turk, 2011a), which is about 1 month after the outside temperature peak. However, what the real proportion is

Natural and Anthropogenic Impacts on Cave Climates
ISBN 978-0-12-822954-5
https://doi.org/10.1016/B978-0-12-822954-5.00019-6

between human-induced heat transmitted into the cave air and warm outside air entering the cave and heating the cave air remains debatable.

The fact that in temperate climates mean cave air temperature reflects mean outside cave temperature (Smithson, 1991) can give a greater insight into the Postojna Cave microclimate. Regarding our air temperature data set, the mean outside cave air temperature for the period 2010–19 was 9.89°C. The mean air temperature for cave monitoring sites is lower than the mean outside value in Črna Jama and at the entrance to Otoška Jama (Fig. 5.1.7.1). The mean air temperature in Lepe Jame, Velika Gora, Sepolcro, Biospeleološka Postaja, Pivka Jama, and Otoška Jama is higher than the mean outside value (Fig. 5.1.7.1). Only at the Pulpito site is the mean temperature of 9.85°C (2017–19) comparable with the mean outside air temperature for the period 2010–19, which is 9.89°C. Cave monitoring sites that have higher temperature values than outside can, at some extent, be connected with an anthropogenic influence. The only exception is the Otoška Jama 2 site, which is not a touristic part of Postojna Cave and is overheated due to natural conditions. At the Otoška Jama 2 site, the mean temperature for the 2018–19 period was 12.51°C, which is 2.62°C higher than outside conditions for 2010–19. In fact, this monitoring site is the most overheated site of all the cave monitoring sites. Increased mean air temperature in cave monitoring sites (Fig. 5.1.7.1), such as in Velika Gora, Koncertna Dvorana, and Lepe Jame, cannot be only due to the presence of visitors, but also an increase in the outside air temperature, especially due to warmer winters in recent years, which causes winter cave air temperatures to be higher than they used to be.

The mean air temperature at the Postojna Cave sites (Postojna 1, Postojna 2, Postojna 3, Koncertna Dvorana) and outside the cave increased in the period from 2010 (or 2011) to 2019 (Fig. 5.1.7.2, Table 5.1.7.1). At Velika Gora (Postojna 1 site) the air temperature increase was the highest, at +0.7°C. The lowest increase of +0.39°C was at Koncertna Dvorana. Also, the mean outside air temperature and visitor numbers have been increasing every year (2010–19).

The artificial opening of the main Postojna Cave entrance in 1866 permitted a more direct exchange between the outside air currents and the most visited cave passages (Stara Jama, Velika Gora, Koncertna Dvorana, Lepe Jame). This is an additional factor for the increase in the outside cave air temperature influencing the increase in the cave air temperature.

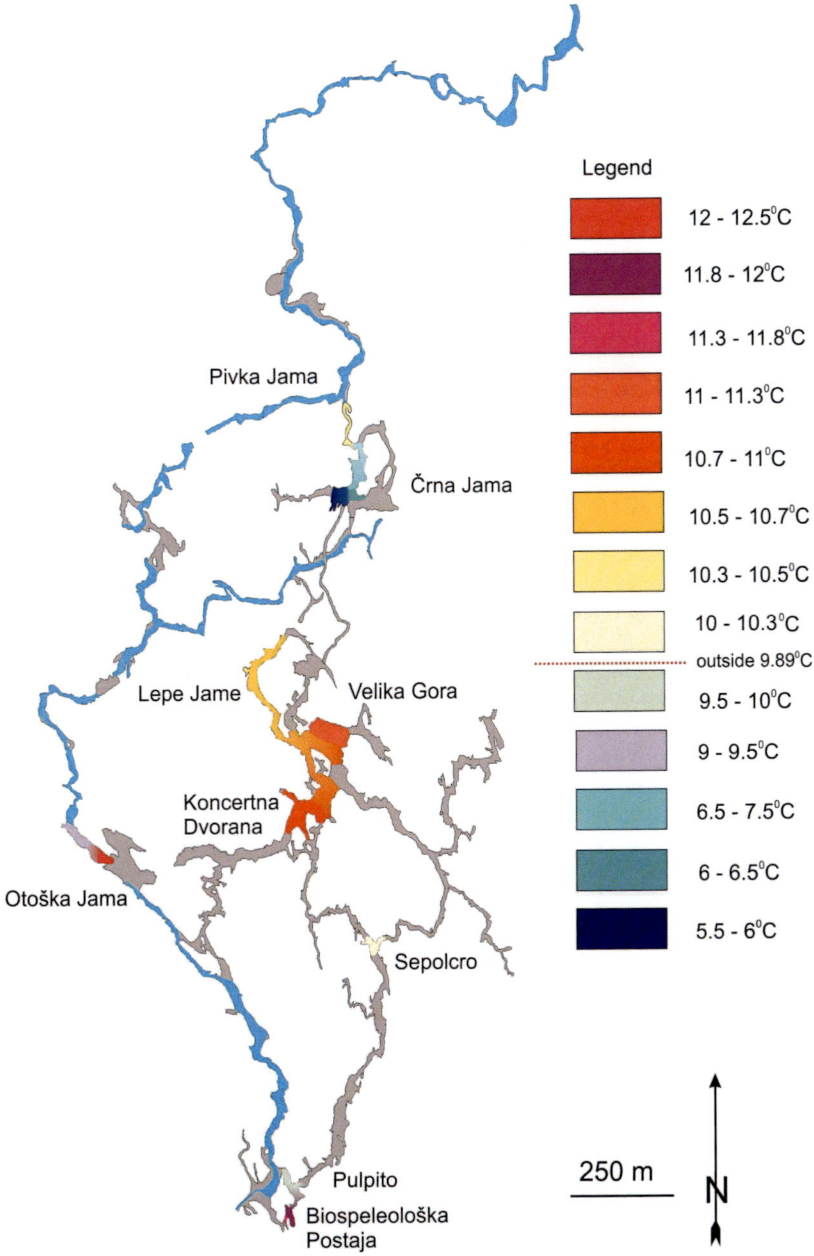

Figure 5.1.7.1 Mean air temperature at Postojna Cave monitoring sites in °C. Legend: 1-Otoška Jama 2 (2018–2019), 2-Biospeleološka Postaja 1 (2016–2019), 3- Biospeleološka Postaja 2 (2016–2019), 4-Velika Gora (Postojna 1, 2010–2019), 5-Koncertna Dvorana (2011–2019), 6-Lepe Jame (Postojna 3, 2010–2019), 7-Lepe Jame (Postojna 2, 2010–2019) and Pivka Jama (Črna Jama 4, 2015–2019), 8-Sepolcro (2016–2019), 9-Pulpito (2017–2019), 10-Otoška Jama 1 (2018–2019), 11-Črna Jama 3 (2015–2019), 12-Črna Jama 1 (2015–2019), 13-Črna Jama 2 (2015–2019).

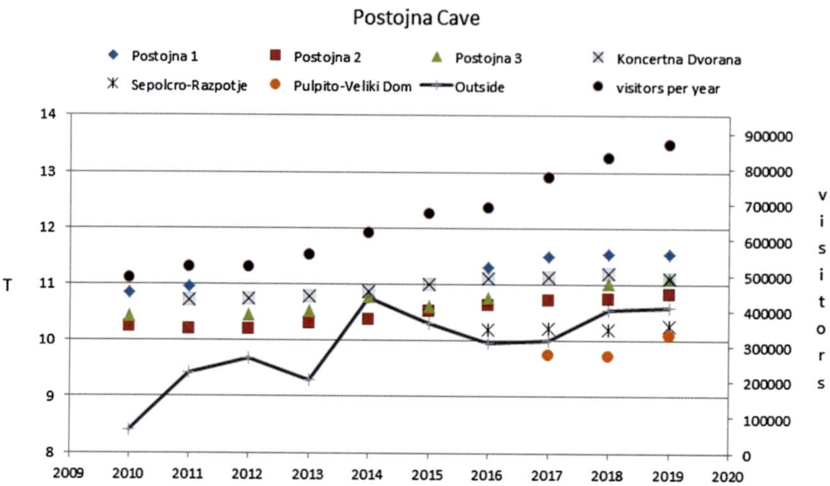

Figure 5.1.7.2 Mean annual air temperature for Postojna Cave monitoring sites and outside with number of visitors per year (2010–2019).

Table 5.1.7.1 Rise of mean annual air temperature (°C) and rise of number of visitors for Postojna Cave (and outside) for the period 2010 (or 2011) to 2019.

	Δ*T*/visitors	**Period**
Postojna 1	+0.7°C	2010–2019
Postojna 2	+0.59°C	2010–2019
Postojna 3	+0.57°C	2010–2019
Koncertna Dvorana	+0.39°C	2011–2019
Outside	+2.18°C	2010–2019
Visitors per year	+377,734	2010–2019

The natural event in winter 2012 should not be forgotten [see Chapter 5.1.3.3, Lepe Jame (Postojna three site)], in which more days of the long northern "bora" wind lowered the air temperature at Postojna 3 site (about 1.5 km deep inside the cave) by about 0.5°C, which is a higher value than the increase in the temperature of 0.3°C during days with high visitor numbers at the same location (Šebela et al., 2015a). Extreme outside cave natural events can have a significant impact on cave climate, and should not be underestimated.

CHAPTER 5.1.7.1

Difference between day and night air temperatures in Postojna Cave

Hourly air temperature monitoring at a number of locations in Postojna Cave since 2009 have shown that mean annual values are increasing (see Fig. 5.1.7.2 in Chapter 5.1.7, Anthropogenic and natural impacts on Postojna Cave microclimate) due to natural and anthropogenic factors. Visits to Postojna Cave are organized during daylight hours and during night hours the cave is closed. To see if the cave air temperature recovers during the night hours we present year-long graphs for 2011 and 2018, comparing measurements at midday (12:00 a.m.) and midnight (0:00 a.m.). During the daytime the cave is heated not only by visitors but also by electric lights, which illuminate the cave passages.

At Velika Gora site (Postojna 1) air temperature measurements at midday were generally slightly higher than midnight temperature measurements (Fig. 5.1.7.1). However, there is only a very small difference (0.04−0.07°C). During colder months there is almost no temperature difference between midday and midnight measurements. These small temperature differences became larger in the second half of April 2011 until the middle of October 2011, which is when the cave is under "summer ventilation" ($T_{out} > T_{cave}$). Meanwhile, during "winter ventilation" ($T_{out} < T_{cave}$) midnight and midday air temperatures are comparable. The period between April and October is also the period with the highest number of visitors. The peak visitor day in 2011 was August 15, with 4700 visitors. The difference between day and night temperature measurements was not significantly different on August 15, 2011, to other summer days (Fig. 5.1.7.1.1).

As has already been described in Chapter 5.1.3.4 [Lepe Jame (Postojna 2 site)], the Postojna 2 site in Lepe Jame passage is a special microclimatic location as illustrated in Fig. 5.1.7.1.2. In summer, this cold-air trap location does not show a difference between day and night air temperature values. However, there are three exceptional events. The first exception was on June 19, 2011, when the midnight value was 10.39°C, while the midday

Natural and Anthropogenic Impacts on Cave Climates
ISBN 978-0-12-822954-5
https://doi.org/10.1016/B978-0-12-822954-5.00020-2

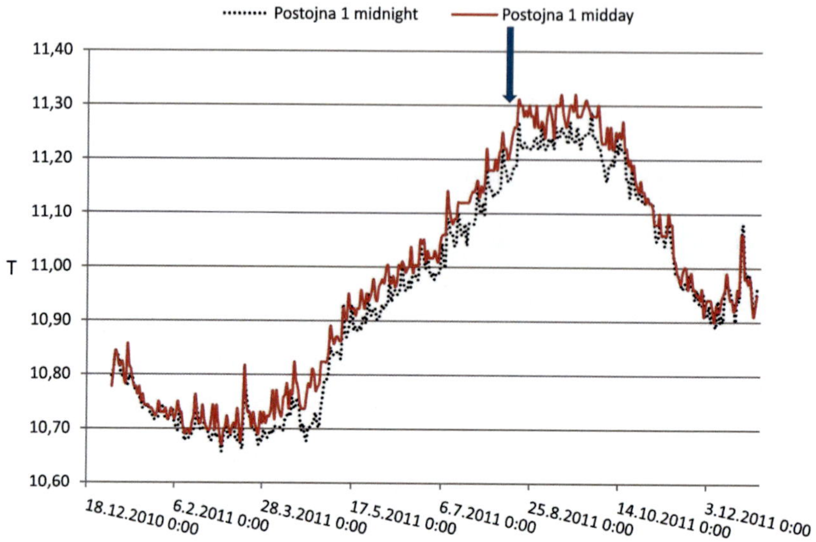

Figure 5.1.7.1.1 Postojna 1 (Velika Gora) in 2011 during midnight and midday measurements in °C. Blue arrow points to August 15, 2011 with the highest number of visitors (4700 visitors).

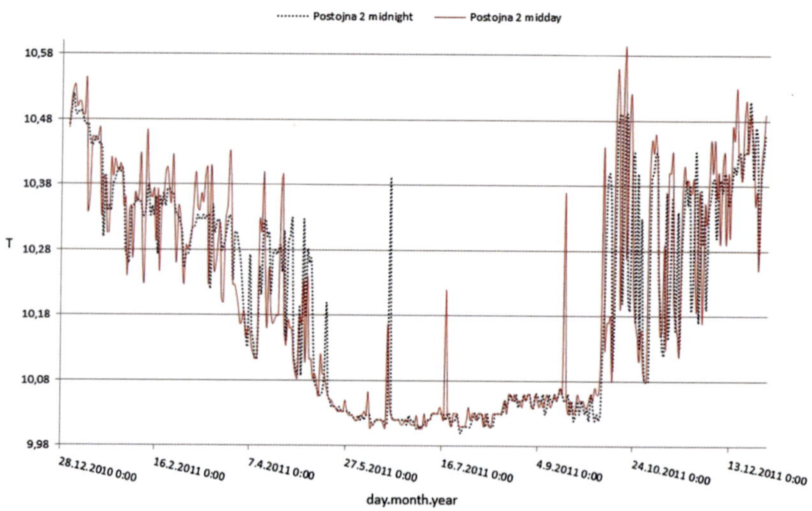

Figure 5.1.7.1.2 Postojna 2 in 2011 during midnight and midday measurements in °C.

value was 10.16°C. On July 18, 2011, during the second exception, the midday temperature value was 10.22°C, while the midnight temperature was 10.02°C. The third exception was on September 18, 2011, when the midday temperature value was 10.37°C, while the midnight temperature was 10.06°C (the difference between midday and midnight was 0.31°C). The three exceptional events in summer 2011 cannot be attributed to the impact of human presence during daylight hours but rather to the natural climate conditions behind the Postojna 2 site (Šebela and Turk, 2011a). Due to the particular characteristics of this monitoring site, the differences between midday and midnight temperature measurements in colder months must also be related to natural climate conditions.

For 2011 we do not have an entire year-long data set of air temperature measurements for the Postojna 3 site in Lepe Jame passage (see Fig. 5.1.7.2 in Chapter 5.1.7, Anthropogenic and natural impacts on Postojna Cave microclimate) due to issues with instruments. This is also the reason that we do not show day and night temperature values for 2011 for the Postojna 3 site.

Koncertna Dvorana (Fig. 5.1.7.1.3) day and night temperature values in 2011 give a similar picture to the Postojna 1 site (Fig. 5.1.7.1.1). Midnight values are generally lower than midday values, especially during warmer months. The maximum difference is recorded as ~0.34°C, as was the case on September 29, 2011, but generally this difference is about 0.15−0.2°C. Also, the most heavily visited day in 2011 (August 15) did not stand out from other August days.

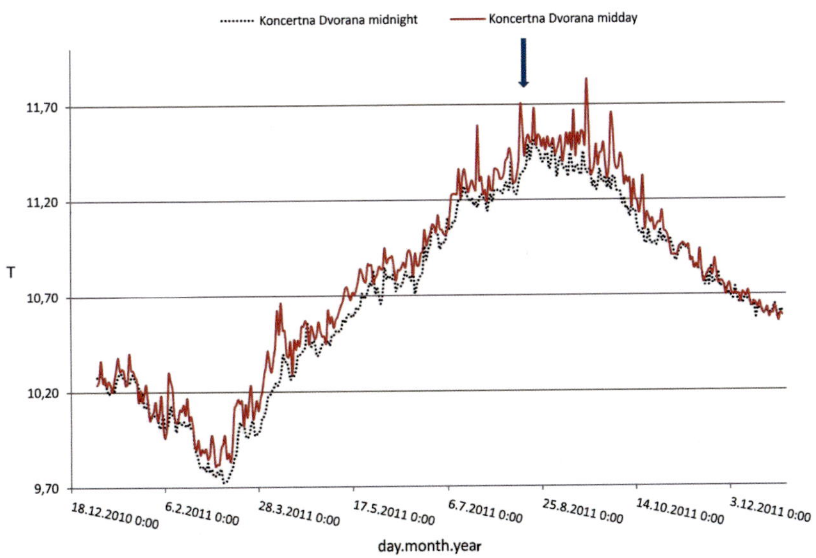

Figure 5.1.7.1.3 Koncertna Dvorana in 2011 during midnight and midday measurements in °C. Blue arrow points to August 15, 2011 with the highest number of visitors (4700 visitors).

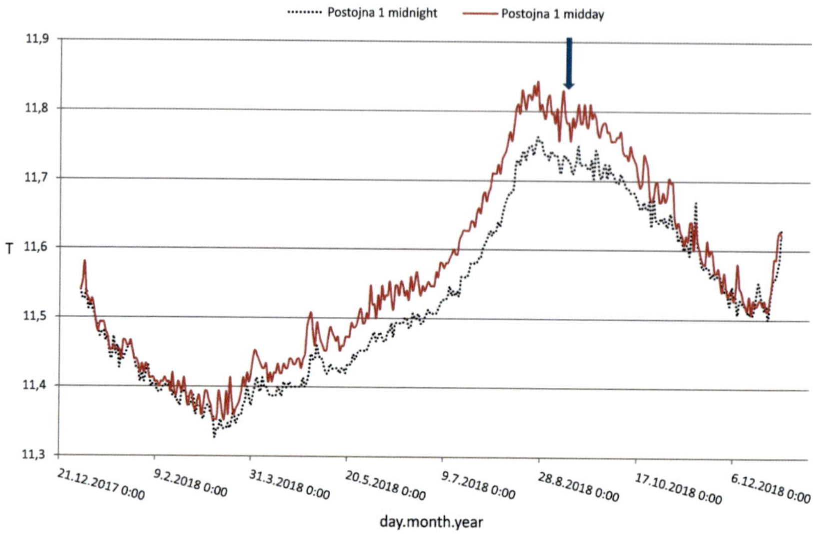

Figure 5.1.7.1.4 Postojna 1 in 2018 during midnight and midday measurements in °C. Blue arrow points to August 15, 2018 with the highest number of visitors (7800 visitors).

Three monitoring sites (Postojna 1, Postojna 2, and Koncertna Dvorana) did not show a large difference between midday and midnight temperatures in 2011.

The day and night temperatures for the 7 years after 2011 are now analyzed. In 2018, the Velika Gora site (Postojna 1) recorded about a 0.07°C difference between day and night values (Fig. 5.1.7.1.4), which is slightly higher but remains very similar to conditions in 2011 (Fig. 5.1.7.1.1). Also, the peak visitor day on August 15, 2018, with 7800 visitors, did not have a significant difference between day and night values. However, between 2011 and 2018 there was a difference in the range of air temperatures. In 2018, at Postojna 1 site the night temperature range was from 11.32°C to 11.76°C, while the daytime temperature range was from 11.35°C to 11.84°C. In 2011, at the Postojna 1 site the night temperature was from 10.66°C to 11.29°C, while the midday values were from 10.67°C to 11.32°C. It can be seen that the midnight temperature values in 2018 at the Velika Gora site were 0.47°C to 0.66°C higher than in 2011. Midday temperatures in 2018 at the Velika Gora site were 0.52°C to 0.68°C higher than in 2011.

The Postojna 2 site day and night temperatures in 2018 (Fig. 5.1.7.1.5) are similar to those from 2011 (Fig. 5.1.7.1.2). However, also here, as was

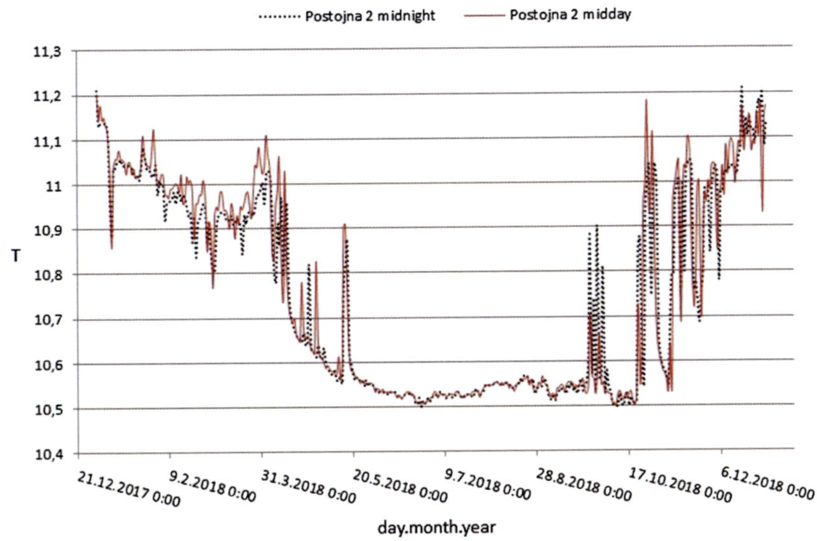

Figure 5.1.7.1.5 Postojna 2 in 2018 during midnight and midday measurements in °C.

the case for Velika Gora (Fig. 5.1.7.1.4), summer values are around 10.55°C, while in summer 2011 they are around 10°C. We can see that the cold air-trap summer conditions at the Postojna 2 location are ~0.5°C higher in 2018 than in 2011.

The Postojna 3 site in Lepe Jame shows a higher temperature difference between midday and midnight values (Fig. 5.1.7.1.6) compared with the Postojna 1 site. The midnight temperature range was 10.67–11.32°C and the midday temperature range was 10.68–11.39°C.

The Koncertna Dvorana site in 2018 (Fig. 5.1.7.1.7) has a midnight temperature range from 10.33°C to 11.9°C, and a midday temperature range from 10.46°C to 12.24°C. In 2011, the same location had a midnight temperature range from 9.73°C to 11.51°C and a midday temperature range from 9.81°C to 11.83°C. The situation for night time shows a temperature difference of 0.39–0.6°C, meaning that midnight air temperatures at Koncertna Dvorana are 0.39–0.6°C higher in 2018 than in 2011. Midday temperatures in 2018 are also higher than in 2011 (by 0.41–0.65°C).

The relation between day and night air temperatures at the Sepolcro site in 2018 (Fig. 5.1.7.1.8) shows a small difference with slightly higher (up to 0.15°C) values for midday measurements, especially during warmer months. The Pulpito site (Fig. 5.1.7.1.9) has special behavior, which is

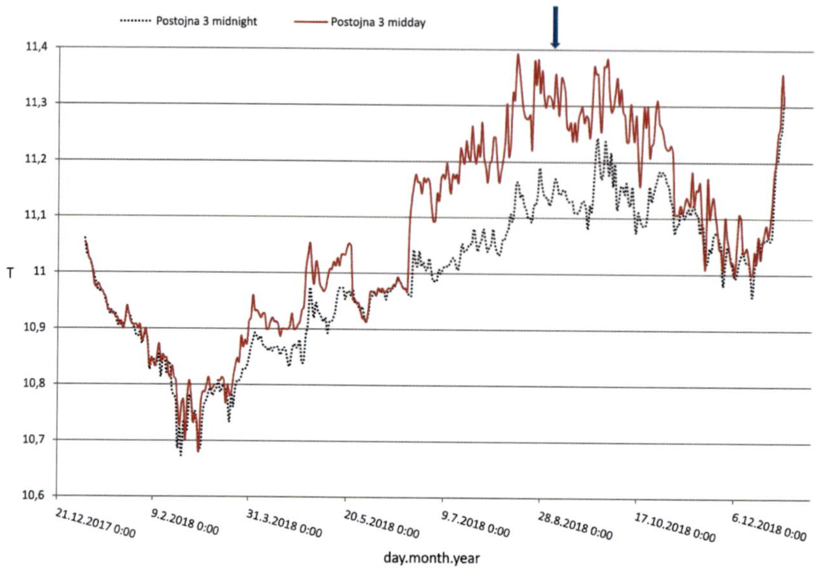

Figure 5.1.7.1.6 Postojna 3 in 2018 during midnight and midday measurements in °C. Blue arrow points to August 15, 2018 with the highest number of visitors (7800 visitors).

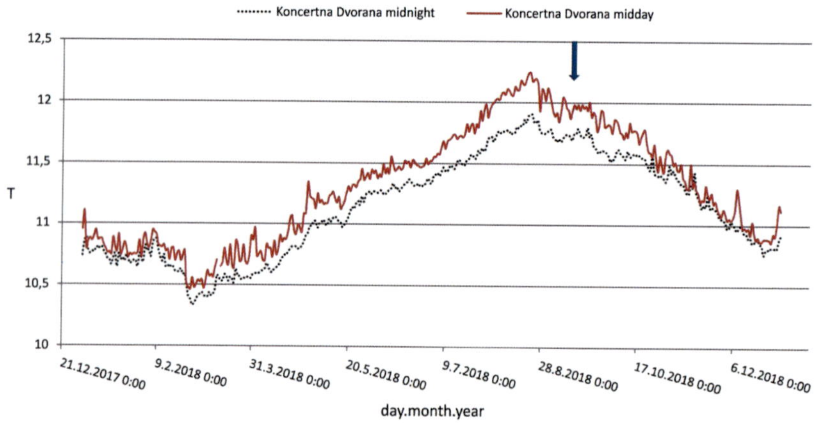

Figure 5.1.7.1.7 Koncertna Dvorana in 2018 during midnight and midday measurements in °C. Blue arrow points to August 15, 2018 with the highest number of visitors (7800 visitors).

during selected periods more similar to the Postojna 2 site (Fig. 5.1.7.1.5). Close to the cave entrance there were almost equal day and night temperature curves during colder months at the Pulpito site, while warmer months show periods when night air temperatures were 2°C higher than

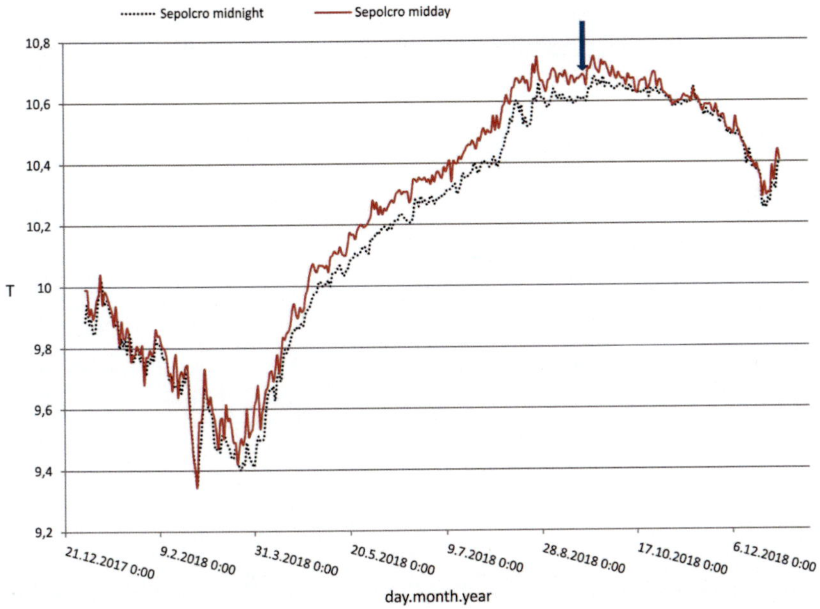

Figure 5.1.7.1.8 Sepolcro in 2018 during midnight and midday measurements in °C. Blue arrow points to August 15, 2018 with the highest number of visitors (7800 visitors).

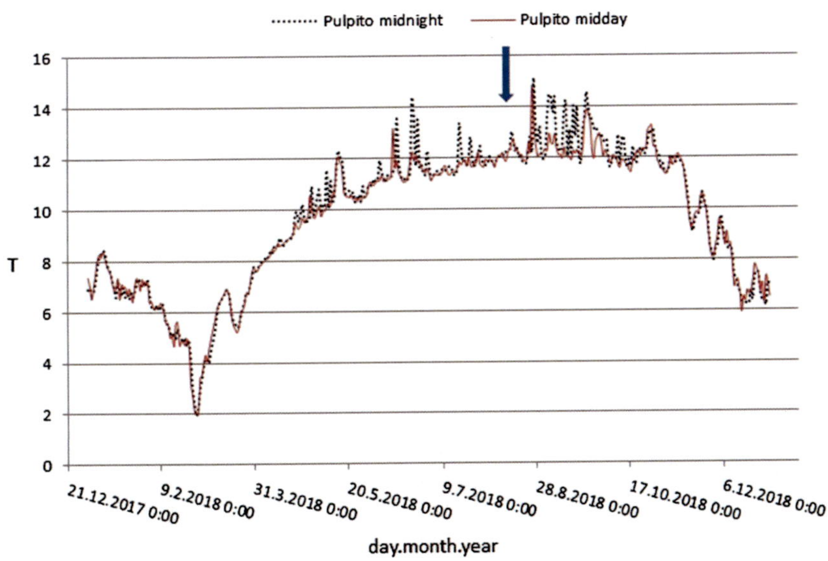

Figure 5.1.7.1.9 Pulpito in 2018 during midnight and midday measurements in °C. Blue arrow points to August 15, 2018 with the highest number of visitors (7800 visitors).

Table 5.1.7.1.1 Pearson correlation coefficients (PCC) for air temperatures and visitors in Postojna Cave for 2011.

	PCC
Postojna 1, midday–midnight	0.99
Postojna 2, midday–midnight	0.87
Koncertna Dvorana, midday–midnight	0.98
Postojna 1, midday–visitors	0.61
Postojna 1 midnight–visitors	0.56
Koncertna Dvorana, midday–visitors/day	0.69
Koncertna Dvorana, midnight–visitors/day	0.66
Postojna 1, midday–outside	0.50
Postojna 2, midday–outside	−0.79
Koncertna Dvorana, midday–outside	0.63

0.90–0.99 (very strong correlation)
0.70–0.89 (strong correlation)
0.40–0.69 (medium correlation)

day temperatures (June, 2018). This shows the significant impact of outside air conditions on the Pulpito site not only during the day time but also at night.

A comparison between midday and midnight air temperatures for Postojna Cave with Pearson correlation coefficients (PCCs) is presented for 2011 (Table 5.1.7.1.1) and 2018 (Table 5.1.7.1.2). A very strong correlation was calculated for the Postojna 1 and Koncertna Dvorana monitoring sites, and a strong correlation was also found for the Postojna 2 site. This shows that the temperature curve for midday data is very well correlated with the midnight data. The relation between midday air temperatures and number of visitors in 2011 has a medium correlation with the Postojna 1 and Koncertna Dvorana sites. A medium correlation is characteristic also for the

Table 5.1.7.1.2 Pearson correlation coefficients (PCC) for air temperatures in Postojna Cave in 2018.

	PCC
Postojna 1, midday–midnight	0.98
Postojna 2, midday–midnight	0.95
Postojna 3, midday–midnight	0.93
Koncertna Dvorana, midday–midnight	0.98
Sepolcro, midday–midnight	0.99
Pulpito, midday–midnight	0.98

0.90–0.99 (very strong correlation)

relation between midnight temperatures and the number of visitors, but with a 0.05 lower PCC value for the Postojna 1 site and a 0.03 lower PCC value for Koncertna Dvorana (Table 5.1.7.1.1). The relation between cave midday air temperatures and outside temperatures shows a medium correlation for Postojna 1 and Koncertna Dvorana and a strong negative correlation for the Postojna 2 site. For 2018 there is not a precise number of visitors, so only relations between midday and midnight temperatures are calculated (Table 5.1.7.1.2). All calculated cave sites (Postojna 1, Postojna 2, Postojna 3, Koncertna Dvorana, Sepolcro, and Pulpito) show a very strong correlation between day and night temperatures. Even the Postojna 2 site midday and midnight temperatures have a higher correlation in 2018 (PCC = 0.95) then in 2011 (PCC = 0.87). This shows that in 2011 there was a strong correlation and in 2018 it was a very strong correlation, and therefore the difference between day and night temperature behavior at the Postojna 2 site was less pronounced in 2018 than it was in 2011.

In previous studies (Šebela and Turk, 2011a; Pipan et al., 2019) a clear 24-h signal and a weaker 12-h signal were determined for the Postojna 3 site (July 24, 2009—June 18, 2010). The Postojna 2 site had an even weaker 24-h signal than the Postojna 3 site and there was an unclear 12-h signal for August 1, 2009—August 18, 2009 (Šebela and Turk, 2011a), which indicated its relatively greater isolation from the rest of the cave (Pipan et al., 2019).

CHAPTER 5.1.7.2

Extreme natural events in Postojna Cave

Contents

During the 200-year modern tourist history of Postojna Cave, the underground and surface landscape have survived different extreme natural events such as floods, earthquakes, and lightning (Šebela 2010, 2011). In addition the area has had to deal with drought, sleet, and strong northern "bora" winds. Such events can sometimes be described as natural disasters and are well documented.

Floods

A large flood in front of Postojna Cave in 1577 destroyed the Modrijan watermill (Habe, 1996, p. 46). This older watermill was placed closer to Pivka ponor than the current Modrijan watermill (Kranjc et al., 2007) (Fig. 5.1.7.2.1), which ceased operation on November 7, 1972, when a high Pivka River carried away the last two mill wheels (Habe, 1996, 52).

In 1931, at Tartarus passage (Fig. 5.1.7.2.2), a 9 m long artificial tunnel was opened for research purposes. In January 1932, an underground research station with horizontal pendulums with photographic recording equipment was placed inside the tunnel. The two pendulums recorded minimum vertical declinations. During this research, rapid growth of the underground River Pivka was detected and lowering of the western mass block was described (Carnera, 1933). Such studies are important for understanding earth tides, underground karst hydrology, and seismology (Carnera, 1933).

Natural and Anthropogenic Impacts on Cave Climates
ISBN 978-0-12-822954-5
https://doi.org/10.1016/B978-0-12-822954-5.00021-4

Figure 5.1.7.2.1 Modrijan watermill marked with arrow in front of Postojna Cave on Pivka River. View towards East, contrary to river flow. *(Photo: S. Šebela.)*

In historic publications (Grotte demaniali di Postumia, 1938) it has been described that in autumn 1933, instruments inside the artificial tunnel were flooded due to underground River Pivka flooding. Instruments were destroyed and removed from the cave. The position of the instruments at the time of the 1933 floods was 519.8 m above sea level (Fig. 5.1.7.2.2). At the end of August 1934, the underground laboratory restarted work, but new floods in December 1934 again interrupted measurements in Tartarusu passage. In 1936, instruments were placed in another part of Postojna Cave, out of the reach of high floods. In 1937, measurements of earth tides continued in Postojna Cave but it is unclear whether this was in Tartarus tunnel or elsewhere (Grotte demaniali di Postumia, 1938).

The high water level in Postojna Cave in 1933 was part of a regional flooding event that took place on September 23–24, 1933, and affected a large part of Slovenia and even part of Croatia (Komac et al., 2008).

As evidenced on the eastern wall of Modrijan watermill (Fig. 5.1.7.2.3), the highest recent flood occurred during the September 19, 2010, event (Fig. 5.1.7.2.4), when water reached 520.4 m (Šebela, 2011). This means that at Pivka River ponor the water level was 9 m higher than normal. In the cave at Tartarus passage the water reached 518 m (Fig. 5.1.7.2.2), which was almost 2 m lower than during the flood event in 1933 (Šebela, 2011).

On November 27, 1987, the water at Modrijan watermill reached 520 m (Fig. 5.1.7.2.3). On December 25, 2009, Pivka River reached 515 m

Figure 5.1.7.2.2 Extreme natural events in Postojna Cave. Legend: a-hight floods (1) in 1933 and (2) in 2010, b-stalagmite fall due to earthquake in 1926, c- stroke of lightning (1) in 2004 and 2006 and (2) in 2008, d-cave air temperature decrease due to "bora" wind.

Figure 5.1.7.2.3 Levels of high floods on Modrijan watermill in front of Postojna Cave. *(Photo: S. Šebela.)*

Figure 5.1.7.2.4 Ponor of Postojna Cave during high floods in September 2010 (left) and during normal water level in March 2013 (right). *(both Photos S. Šebela.)*

above sea level and on December 12, 2008, the water was up to 511.3 m above sea level (Fig. 5.1.7.2.3).

During heavy precipitation on December 25, 2009, at Stara Jama passage percolating water from the surface to the cave caused parts of the cave train rails to be flooded, which was not due to the rise in the underground River Pivka level. This event caused some issues with visitor transport to the cave.

With reference to historic and actual data it is clear that the high flood event in 1933 represents the highest water level in Postojna Cave over the last 100 years.

Drought

The cold winter of early 2012 was followed by a warm and dry summer. The Pivka River bed was almost completely dry (Fig. 5.1.7.2.5). Also, the cave entrance chamber at Veliki Dom was dry and it was possible to walk on the bottom of the chamber. During this dry period, 970 m^3 of fluvial sediments, tree branches, and other materials dumped at Velik Dom by Pivka River were removed in a cleaning operation. The previous cleaning activity in Velik Dom had taken place in 1983. This means that between 1983 and 2012 there were no years in with such dry conditions suitable for cleaning had occurred.

Earthquakes

It is generally claimed that caves are stable environments during earthquakes. However, there is some good proof for active tectonic

Figure 5.1.7.2.5 Pivka River bed on August 27, 2012 in front of Postojna Cave. *(Photo: S. Šebela.)*

deformations in caves (Gilli, 1999; Postpischl et al., 1991; Becker et al., 2005; Kagan et al., 2005; Šebela, 2008, 2010).

In Postojna Cave during earthquakes the sound can be similar to an approaching train, with explosions and movements of the ground having been described (Glažar, 2006; Šebela, 2008, 2010). In Biospeleološka Postaja, which is situated close to the cave entrance, cave guides described shaking of the ground and fence.

One of the strongest earthquakes in Slovenia (Idrija Earthquake in 1511), and also the Ljubljana earthquake in 1895, have no available written testimonies about possible damage to Postojna Cave as a result. Probably the oldest description belongs to Schmidl (1854) who described the February 2, 1834 earthquake, which was strongly felt in the area of Postojna. This was an earthquake with a magnitude 3.9 and brontide sound in the vicinity of Postojna (Javorniki Mountains) (Ribarič, 1982).

The so-called Cerknica earthquake that was strongly felt in Postojna occurred on January 1, 1926, with a magnitude of 5.6 (Ribarič, 1982; Poljak et al., 2000; Šebela, 2010). The *Edinost* newspaper published that inside the Postojna Cave "beautiful stalagmite, which measured almost 1 m in diameter, collapsed" (Edinost, January 8, 1926, 8). According to Ribarič (1982) this earthquake was located at the Javorniki Mountains but closer to Cerknica than Postojna. Zanon (1926) described the January 1, 1926 event as the Venice earthquake with its location closer to Postojna than Cerknica. In Zanon's (1926) article, according to testimony of Ivan Andrej Perko, who was the director of Postojna Cave, inside Postojna Cave there was no damage but an explosion that accompanied the earthquake was heard. Brontide sounds and lightning accompanying the earthquake were heard and seen as far away as Venice, Italy (Zanon, 1926).

However, in the annual report of the management of Postojna Cave (Grotte demaniali di Postumia, 1928) for the period July 1, 1926 to December 31, 1927, on page 12, it is written in Italian: "*Sgombro dei pezzi di una grande colonna rovesciatasi nella Scala del Candore in seguito al terremoto del giorno 1 gennaio 1926*" (translation: "Removal of large pieces of column that collapsed at Scala del Candore as a consequence of the earthquake on January 1, 1926"). The note in the newspaper (Edinost January 8, 1926, p. 8) and the report of the cave management (Grotte demaniali di Postumia, 1928) confirm the damage to the stalagmites in the chamber, which is situated about 600 m deep inside the cave (Fig. 5.1.7.2.2).

Lightning strikes

Lightning strikes on May 7, 2004, inside Postojna Cave were described by Glažar (2006). During a rain storm, a group of visitors to Postojna Cave were at the base of Velika Gora chamber. They heard an explosion. At first it was completely dark in the cave, than the ceiling of Velika Gora chamber was illuminated. The chamber arch glowed like a spider net of numerous threads in a light blue to green light, which lasted a few seconds (Glažar, 2006). There was a lightning strike on the surface above the cave which passed through the 65 m thick limestone roof to Velika Gora chamber (Fig. 5.1.7.2.2).

Additionally, on May 16, 2006, a barasol instrument for measurement of radon concentrations in the cave atmosphere was found to have been destroyed on the top of Velika Gora collapsed chamber. There was no other explanation for the damage than a lightning strike.

There was another case for a lightning strike at Veliki Dom (Fig. 5.1.7.2.2). In 2008, an instrument installed in the water for conductivity measurements was damaged due to lightning.

Bora wind

As has already been described in Chapter 5.1.3.3, in February 2012 a strong outside northern wind called "bora" with a maximum wind speed of 18 m/s and a mean speed of 3.8 m/s caused a significant temperature change in Lepe Jame at the Postojna 3 monitoring site (Fig. 5.1.7.2.2). Cold air traveled close to the cave floor where it displaced relatively warmer cave air toward the upper parts of the passages and chambers. Natural drivers such as bora wind in winter and other colder times of the year caused a cave air temperature decrease of 0.3°C on February 10, 2012 (Šebela et al., 2015a).

Sleet

Sleet is rain that contains some ice and is formed during special climatic conditions when warmer and colder air masses interact, resulting in icy rain. Such an event happened in Postojna and other parts of Slovenia in January–February 2014. Branches of trees were weighed down with up to 8 cm of ice; forests were destroyed. Above and close to Postojna Cave,

Figure 5.1.7.2.6 The road from Postojna Cave to Postojna town with destroyed trees due to sleet in 2014. *(Photo: S. Šebela.)*

numerous trees were damaged (Fig. 5.1.7.2.6), dramatically changing the karst landscape and vegetation above the cave. The resulting reduction in the number of trees above the cave will have impacted the soil characteristics, probably impacting the dynamics of waters percolating from the surface to the cave.

In the newspaper *Dom in Svet* there was an article about a sleet event, which took place on December 14—15, 1899, in the area of southwest Slovenia (Domicelj, 1900). In the evening of December 14, 1899, the snow changed to rain when the relatively warmer southern wind started to blow. However, after the night the colder northern wind started to blow at lower atmospheric layers (Domicelj, 1900), causing the rain to become icy. Due to the amount of ice on trees, many areas of forest were destroyed. After more than 100 years, sleet returned as an extreme natural event in the area of Postojna.

CHAPTER 5.2

Predjama Cave

Predjama Cave was named after the small village Predjama (translation: in front of the cave), which is situated 10 km northwest of Postojna town. In 2020 the cave officially reached 16,755 m of known passages and became the third longest cave in Slovenia (Cave Register). Speleological research is still on-going, showing that Predjama Cave has great potential for the discovery of new extensive passages. In the first half of 2020 additional new passages were discovered in Zahodni Rov, so that the cave has more than 16 km of passages (Fidel, 2020). In a speleological sense we should describe it as the Predjama Cave system, because there are other caves (Predjama, Požiralnik Lokva, Jama v Grapi, Bojanovo Brezno pri Predjami) composing part of this cave system (Fig. 5.2.1).

The schematic longitudinal section AB (Fig. 5.2.1) demonstrates the complex arrangement of cave passages at more levels, showing entanglement between horizontal dry passages, vertical dry passages, collapsed chambers, and horizontal water passages. Lokva stream sinks from impermeable Eocene flysch rocks into the karst system at 462 m above sea level. Active underground water flow forms passages of Požiralnik Lokva. Konjski Hlev is the middle horizontal level starting at 490 m above sea level. The entrance to Konjski Hlev is partially blocked with a constructed wall and doors (Fig. 5.2.2), however this does not prevent the air movement between the cave and outside. The entrance to Konjski Hlev is basically an entrance to the most extensive know parts of the cave system. From the first collapsed chamber Velika Dvorana, the tourist path leads to Fiženca representing the upper passages and tourist exit. In a speleological sense the middle level continues from Velika Dvorana through Dvorana Dvojčkov and Vetrovna Luknja to Črna Dvorana collapsed chamber. This last chamber is the crossing between Vzhodni and Zahodni Rov (the eastern and western passages). Jama v Grapi on the eastern side of Vzhodni Rov became part of the cave system with a diving connection. Bojanovo Brezno is also part of the Predjama Cave system. This is a system of mostly vertical shafts that join Predjama Cave in Zahodni Rov at the surface.

Predjama is known for its archeological remains, history related to Predjama castle (Fig. 5.2.2), and numerous old inscriptions on the cave walls

Natural and Anthropogenic Impacts on Cave Climates
ISBN 978-0-12-822954-5
https://doi.org/10.1016/B978-0-12-822954-5.00022-6

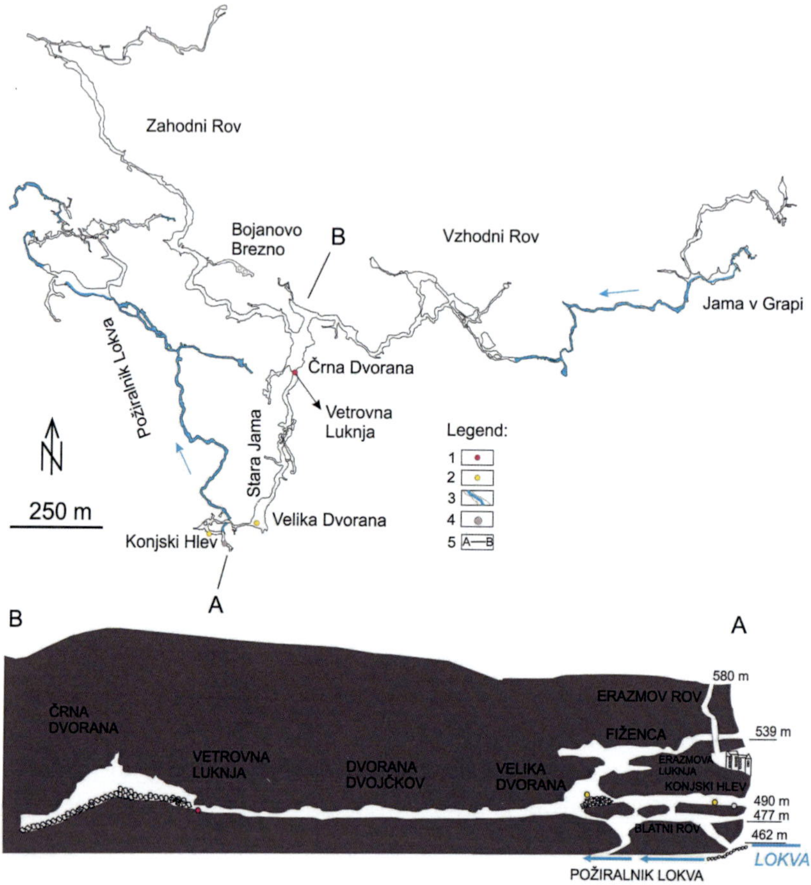

Figure 5.2.1 Predjama Cave System with micro-climatic monitoring sites. Legend: 1-carbon dioxide monitoring, 2-air temperature and air pressure monitoring, 3-water passage. 4-collapse due to high floods in 2010, 5-schematic longitudinal section AB.

(Fig. 5.2.3). In a geological sense it is very interesting because its passages start at the contact between Eocene flysch and Upper Cretaceous limestone and continue through Upper Triassic dolomite, and Jurassic limestone and dolomite (Habe, 1970; Šebela and Čar, 1991; Čar and Šebela, 2001).

The Lokva stream runs over Eocene flysch and sinks at the tectonic contact with Upper Cretaceous limestone (Čar and Šebela, 2001). The underground Lokva resurges 15 km to the northwest in Vipava springs, what was been foreseen by Valvasor (1689). Lokva stream and Vipava River belong to the Adriatic Sea watershed, while the underground Pivka in Postojna Cave belongs to the Black Sea watershed. The drainage divide

Figure 5.2.2 Predjama castle with Predjama Cave entrances. *(Photo: S. Šebela.)*

Figure 5.2.3 Old inscription (1694) on Predjama Cave walls. *(Photo: S. Šebela.)*

between the Adriatic and Black seas is between Predjama and Postojna Caves in the northern Pivka valley.

Predjama Cave is closed for visitors every year during the period of bat hibernation (October–May). In recent years there have been about 1000 bats hibernating in the Predjama Cave during winter (Šebela and Turk, 2014b). Predjama is a major tourist attraction because of the spectacular cave position in the limestone wall and the appearance of its medieval castle (Šebela and Turk, 2014b). The oldest known archeological remains in the entrance parts of Predjama Cave date back to the Eneolithic period with some Neolithic elements (Korošec, 1975), and are closely connected to early human settlement.

The rich cultural layers and numerous discovered objects indicate that Konjski Hlev was safe and suitable for human habitation (Bavdek and Čuk, 2015). People found refuge in the vicinity of the cave in the Copper Age or Eneolithic (Bavdek and Čuk, 2015), and in the Bronze and Iron ages (from the end of the third millennium to roughly the first century BC) the cave served as a temporary hidden refuge or as a site for burying the dead (Bavdek and Čuk, 2015).

Erazmova Luknja is part of a cave system behind the castle (Fig. 5.2.1). A free-standing palatium or palace covered with a four-sided roof was erected in the entrance to Erazmova Luknja in the 12th century (Bavdek and Čuk, 2015). The best know castle administrator was Erasmus Lueger, who headed the castle and the cave until the end of the 14th century (Bavdek and Čuk, 2015). Erasmus (or Erazem in Slovene) was a hero to many, a robber baron, and a member of a wealthy and noble family (Bavdek and Čuk, 2015), who used the vertical Erazmov Rov (Fig. 5.2.1) as a secret passage to escape from the castle.

During the renovation works in 1991 at Predjama castle a treasure cache from the last quarter of the 16th or the early 17th century was found in the outer cellar wall (Bavdek and Čuk, 2015). This showed the important intertwining of the natural cave walls and the human-made castle walls.

Mineralogical studies and dating of thin black deposits on speleothems in Črna Dvorana (meaning Black Chamber) show that they date back to 4095 ± 25 years B.P. (Eneolithic) and were caused by burning fires at the cave entrance during winter (Šebela et al., 2017), as supported by archeological remains (Korošec, 1975).

The Predjama Cave was always interesting for cave meteorology studies due to its complex geometry of passages and the many cave entrances at different levels. Anelli (1941−44) presented meteorological observations in Predjama Cave, including air circulation and temperature measurements. At Konjski Hlev passage, Anelli (1941−44) described two types of air currents with different temperatures. While the first air current entered the cave the second one left. In cold winters, as in January, 1943, relatively warm cave air was exiting of Fiženca Cave (upper part of Predjama Cave), causing a cloud in the early morning hours at the Fiženca Cave entrance (Anelli, 1941−44).

Habe (1970) published data regarding cave meteorology with an air temperature diagram from February 1956 to February 1957 for Lokva, Stara Jama, Fiženca, Blatni Rov, Velika Dvorana, Črna Dvorana, and Vzhodni Rov passages. Additionally, measurements of air currents, humidity, and temperature of the cave rock were presented by Habe (1970). The highest air current speed of 8.3 m/s was measured in February 1956 at Vetrovna Luknja (Habe, 1970). Winter and summer ventilation regimes were identified. During cold winters, ice stalagmites have been found in Dvorana Dvojčkov chamber at depths of 400 m inside the cave, illustrating the intrusion of cold outside air deep into the interior of the cave (Habe, 1970).

Kranjc (1983) described the influence of frost on cave formation and speleothems in the entrance parts of Predjama Cave. Most speleothems were broken in late winter. Besides air temperature and humidity measurements at Predjama Cave entrance, air current speed measurements at Vetrovna Luknja showed the highest value of 9 m/s on January 9, 1981 (Kranjc, 1983).

Measurements of continuous air temperature and air pressure monitoring started on August 6, 2009 in Predjama Cave (Šebela and Turk, 2011b). Two monitoring locations were established. The first monitoring location was set in the entrance part of the cave, in Konjski Hlev, at 488 m above sea level. The second monitoring location was situated in Velika Dvorana (492.5 m). The datalogger was installed 0.3 m above the floor and 0.05 m from the wall in the Konski Hlev site, while it was 0.3 m above the floor and 0.05 m from the limestone block in Velika Dvorana. It was found that the air temperature in Velika Dvorana was much more stable than in Konjski Hlev, which is more subjected to external influences (Šebela and Turk, 2011b). Air can easily enter the cave due to the distribution of great entrances at different altitudes. Also for this reason, Predjama Cave is very dynamic and air penetrates deep into the cave, to at least Vetrovna Luknja (Šebela and Turk, 2011b).

The relationship between visitors to Predjama castle and Predjama Cave, and characteristics of Predjama Cave air temperature monitoring (2009–12) have been studied by Šebela and Turk (2014b). Winter and summer ventilation regimes have been confirmed. The Velika Dvorana monitoring site was presented as a special location where the summer air temperature is stabilized at 9–10°C regardless of the outside cave conditions (Šebela and Turk, 2014b).

In Predjama Cave, the air circulation (speed and direction) was occasionally measured with anemometers (Šebela and Turk, 2014b); Šebela et al., 2017). Not only winter and summer ventilation regimes are presented but also short-term periods with transitions from winter to summer ventilation on a daily basis. Such is the case when during the night the outside air temperature is lower than the cave air temperature and during the day the outside air temperature is higher than the cave air temperature (Šebela and Turk, 2014b).

The most modern studies of Predjama Cave microclimate have started to give a better understand of the influence of the use of these natural treasures for tourism from May to October each year (Šebela and Turk, 2014b). Basic monitoring of air circulation, air temperature, air pressure, and CO_2 has been organized since August 6, 2009, in the Predjama Cave. At one point the microclimatic monitoring was due to be stopped because the anthropogenic impact on the cave microclimate was not confirmed, however it was decided to persevere with the monitoring because this cave can be a reference cave for pure natural conditions. The result of monitoring perseverance is an 11-year air temperature data set.

CHAPTER 5.2.1

Air temperature monitoring

Air temperature monitoring in Predjama Cave started at hourly intervals on August 6, 2009, and continues to this day. The purpose of this monitoring was the organization of expert control of the use of the cave for tourism. Because only entrance parts of the cave are used for tourism and because the number of visitors is low, two air temperature monitoring sites give sufficient information about the thermal characteristics of the entrance areas (see Fig. 5.2.1 in Chapter 5.2, Predjama Cave). Konjski Hlev monitoring site (Fig. 5.2.1.1) is about 70 m from the cave entrance (Fig. 5.2.1.2). The second air temperature monitoring site is installed in the first large collapsed chamber Velika Dvorana (Fig. 5.2.1.3). From Velika Dvorana, the tourist path follows metal ladder steps to reach the upper Fiženca passage. Both temperature instruments are installed some meters away from the tourist path. In Predjama Cave there is no permanent electricity, with visitors using helmets with lights (Fig. 5.2.1.3).

The first systematic air temperature monitoring in Predjama Cave was performed by Anelli (1941–44). In this publication (Anelli, 1941–44), not only were new measurements published but also some historic observations were described. Issues with torches extinguished due to the strong ventilation in Predjama Cave were described by Valvasor (1689), Nagel (1748),

Figure 5.2.1.1 Konjski Hlev in Predjama Cave. White circle points to air temperature monitoring site, view towards South (towards cave entrance), tourist path is 1.5 m wide. *(Photo: S. Šebela.)*

Natural and Anthropogenic Impacts on Cave Climates
ISBN 978-0-12-822954-5
https://doi.org/10.1016/B978-0-12-822954-5.00023-8

Figure 5.2.1.2 Entrance to Konjski Hlev (horse stable) at 490 m above the sea in Predjama Cave. *(Photo: S. Šebela.)*

Figure 5.2.1.3 Velika Dvorana in Predjama Cave. White circle points to air temperature monitoring site, view towards southwest. Ladder to upper passages in Fiženca. *(Photo: S. Šebela.)*

Steinberg (1758), and Schmidl (1854). Nagel's visit to Predjama is documented by his historic inscription on the cave wall (Fig. 5.2.1.4). J. A. Nagel was a mathematician of the Vienna Court and also the creator of the first Postojna Cave map (Kranjc et al., 2007).

The importance of Anelli's (1941–44) studies in Predjama Cave are related to an understanding of different ventilation regimes not only at cave entrances but also deeper inside the cave all the way to Vetrovna Luknja

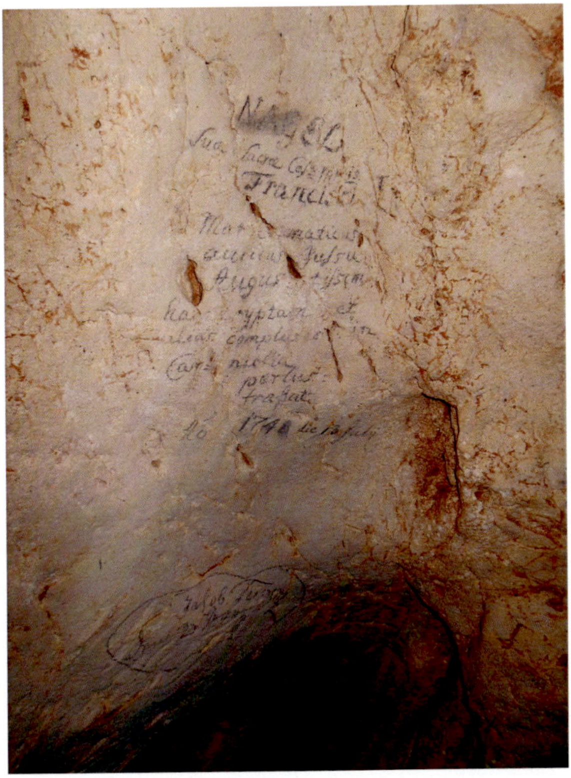

Figure 5.2.1.4 Wall inscription from 1748 of J. A. Nagel in Predjama Cave. *(Photo: S. Šebela.)*

(see Fig. 5.2.1 in Chapter 5.2, Predjama Cave). He also determined "warm cave parts" as Fiženca and "colder cave parts" as ponor of Lokva (Anelli, 1941–44). Winter and summer ventilation regimes determined by Anelli (1941–44) and used also by Habe (1970) were an important base for the most recent studies also (Šebela and Turk, 2011b, 2014b).

An almost 11-year air temperature data set (2009–20) for Predjama Cave is presented in Fig. 5.2.1.5. In a general sense we can see that temperature behavior in Konjski Hlev is similar to the outside cave conditions. At Velika Dvorana chamber, the summer temperature values are not as high as at Konjski Hlev, and do not exceed 9–10°C (Šebela and Turk, 2014b).

If we take a closer look at the measurements for 2019 (Table 5.2.1.1) we can see that the mean annual air temperatures in Konjski Hlev (9.71°C) and Velika Dvorana (7.55°C) are both lower than the mean outside air

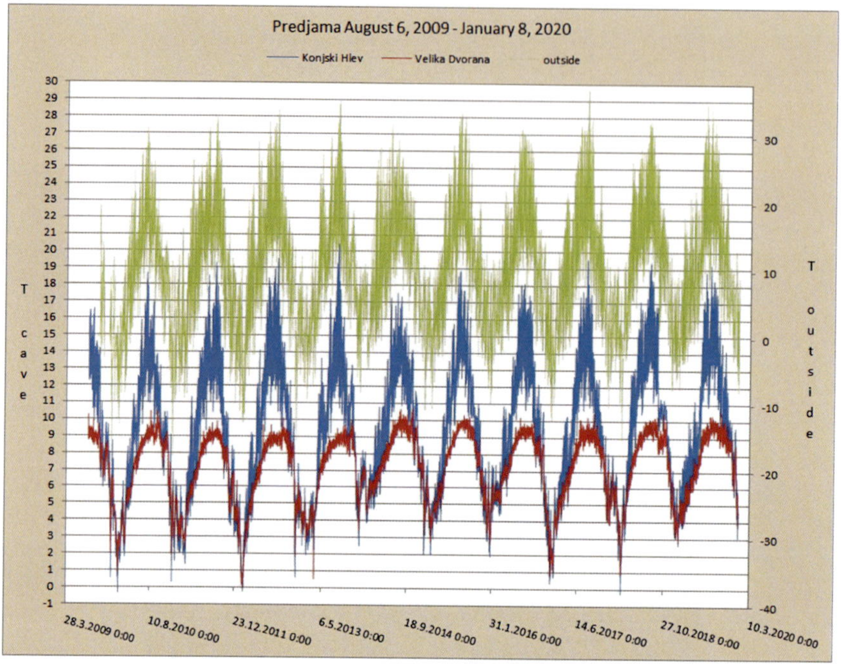

Figure 5.2.1.5 Air temperature dataset for Predjama Cave and outside in °C for August 6, 2009 to January 8, 2020.

Table 5.2.1.1 Basic statistics for 2019 in Predjama Cave in °C.

	Konjski Hlev	Velika Dvorana	Outside
Mean	9.71	7.55	10.58
Max	19.45	10.5	34.92
Min	2.62	2.76	−7.51
Range	16.83	7.74	42.43

temperature. The temperature range over a 1-year period (2019) was 16.83°C for Konjski Hlev and 7.74°C for Velika Dvorana. Both values show the impact of outside climate conditions on cave monitoring sites. This is additionally confirmed with Pearson correlation coefficients (PCC) (Table 5.2.1.2). PCC for the year 2019 is very strong for Konjski Hlev (PCC = 0.9) but not as strong for Velika Dvorana (PCC = 0.74). Velika Dvorana points to special microclimatic conditions, which are not as heavily dependent on the outside climate as in Konjski Hlev.

In 2011, the situation (Table 5.2.1.3) was different to 2019 (Table 5.2.1.1) because the mean annual air temperature was lower in 2011

Table 5.2.1.2 Pearson correlation coefficients (PCC) for air temperatures in Predjama Cave for 2019.

	PCC
Konjski Hlev—Velika Dvorana	0.92
Konjski Hlev—outside	0.9
Velika Dvorana—outside	0.74

0.90—0.99 (very strong correlation)
0.70—0.89 (strong correlation)
0.40—0.69 (medium correlation)

Table 5.2.1.3 Basic statistics for 2011 in Predjama Cave in °C.

	Konjski Hlev	Velika Dvorana	Outside
Mean	8.66	6.67	9.43
Max	19.22	9.87	32.95
Min	1.30	2.20	−10.17
Range	17.92	7.67	43.12

than in 2019. Differences in PCC calculations between 2011 (Table 5.2.1.4) and 2019 (Table 5.2.1.2) were insignificant. The highest difference was in the mean annual air temperature. In 2011, the mean annual air temperature in Konjski Hlev was 8.66°C, while it was 9.71°C in 2019. In 2011, the mean annual air temperature in Velika Dvorana was 6.67°C, and 7.55°C in 2019. This shows that the mean annual air temperature in Konjski Hlev was 1.05°C higher in 2019 than in 2011. In Velika Dvorana, the mean annual air temperature in 2019 was 0.88°C higher than in 2011. Here, we need to stress that the number of visitors in Predjama Cave is less than 10,000 per year and that the human impact on the Predjama Cave air temperature is negligible (Šebela and Turk 2014b). The increase in mean annual air temperature in the Konjski Hlev and Velika Dvorana monitoring sites must be related to outside cave climate conditions. In fact, the mean annual outside cave air temperature was 1.15°C higher in 2019 than in 2011.

Table 5.2.1.4 Pearson correlation coefficients (PCC) for air temperatures in Predjama Cave for 2011.

	PCC
Konjski Hlev—Velika Dvorana	0.94
Konjski Hlev—outside	0.91
Velika Dvorana—outside	0.79

We need to look closer at the conditions for the annual and daily levels (Fig. 5.2.1.6). Daily air temperature oscillations are higher in Konjski Hlev than in Velika Dvorana, especially during the warmer times of the year when they can reach 3.7°C. Daily temperature oscillations in Velika Dvorana over the whole year are much lower than in Konjski Hlev. Daily oscillations in Velika Dvorana are generally lower than 1°C. At the colder times of the year ($T_{cave} > T_{out}$) air temperature values between Konjski Hlev and Velika Dvorana are more comparable, than in warmer periods ($T_{cave} < T_{out}$).

A comparison of the measured Predjama Cave temperature values with the outside air temperatures has already been published for the period from August 6, 2009 to December 31, 2012 (Šebela and Turk, 2014b). It was found that cave air temperatures at Konjski Hlev increase with air temperature increases outside the cave (Šebela and Turk, 2014b). At Velika Dvorana, the cave air temperature increases but when the outside temperature exceeds 15°C the cave air temperature at Velika Dvorana is stabilized (Šebela and Turk, 2014b). The Velika Dvorana summer air temperature limit of 9–10°C is similar to the mean annual air temperature outside the cave (Šebela and Turk, 2014b).

Relatively cold and wet winter air is heavier than warmer and generally drier summer air. During the winter period cold and wet air can readily break through into the deeper parts of the cave passages (Šebela and Turk,

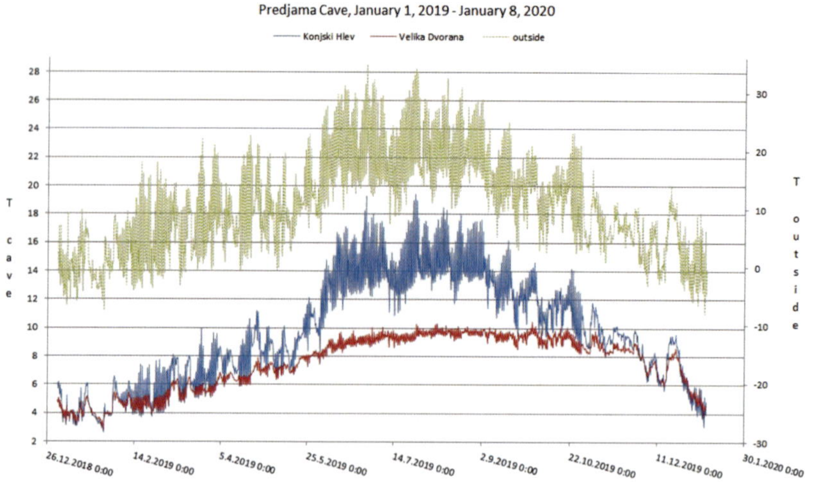

Figure 5.2.1.6 Hourly air temperature characteristics in Predjama Cave and outside the cave in °C for 2019.

2014b). In colder periods of the year temperature values in Konjski Hlev and Velika Dvorana drop due to winter ventilation or cold outside air entering the cave (Fig. 5.2.1.6).

External temperature influences on monitoring locations within the cave also can be studied with spectral analysis (Bourges et al., 2006). Spectral analysis performed for the time period December 2009–January 2010 showed significant 12 and 24 h periodicity for both Predjama Cave locations (Šebela and Turk, 2014b). The spectral density was higher in Konjski Hlev (which is closer to the entrance) than in Velika Dvorana (which is farther from the entrance). Similar results with spectral analysis were shown for the Postojna Cave monitoring sites (Pipan et al., 2019).

When outside air temperatures are very warm during the day and are below cave air temperatures at night significant daily oscillations at both cave monitoring sites were obtained (Fig. 5.2.1.7). Intrusion of cold night outside air into the cave appears to be the same in Konjski Hlev and Velika

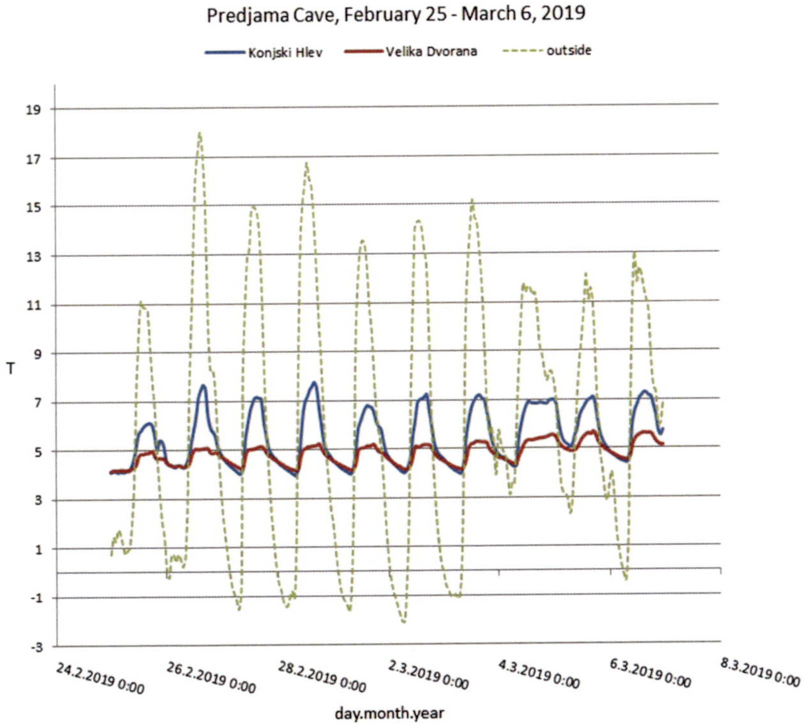

Figure 5.2.1.7 Hourly characteristics of air temperature in °C in Predjama Cave and outside for February 25 – March 6, 2019.

Dvorana because the minimum night temperatures are synchronized at both locations. Maximum daily temperature values are higher close to the cave entrance in Konjski Hlev than in Velika Dvorana.

Summer characteristics when T_{out} is higher than T_{cave} (Fig. 5.2.1.8) show stable air temperature in Velika Dvorana, as already described by Šebela and Turk (2014b). This situation is somehow comparable with Postojna 2 in Lepe Jame (Postojna Cave) summer conditions [see Chapter 5.1.3.4, Lepe Jame (Postojna 2 site)]. As at the Postojna 2 site, at the Velika Dvorana site upward airflow mode (UAF) during winter (Faimon et al., 2012) is pronounced, while summer conditions show stabilized air temperature at both locations (Postojna 2 in Postojna Cave and Velika Dvorana in Predjama Cave) point to a stable temperature situation. At the Postojna 2 site summer daily oscillations are negligible and for 3 months were stabilized at 10.6°C, but at Velika Dvorana daily oscillations can reach 0.6°C and are stabilized at around 9–10°C. When air temperature on the surface exceeds 25°C the cave air temperature at Velika Dvorana remains constant at around 9–10°C (Šebela and Turk, 2014b). In summer the cave air remains as wet and heavy as winter air on the surface, and the warmer summer air

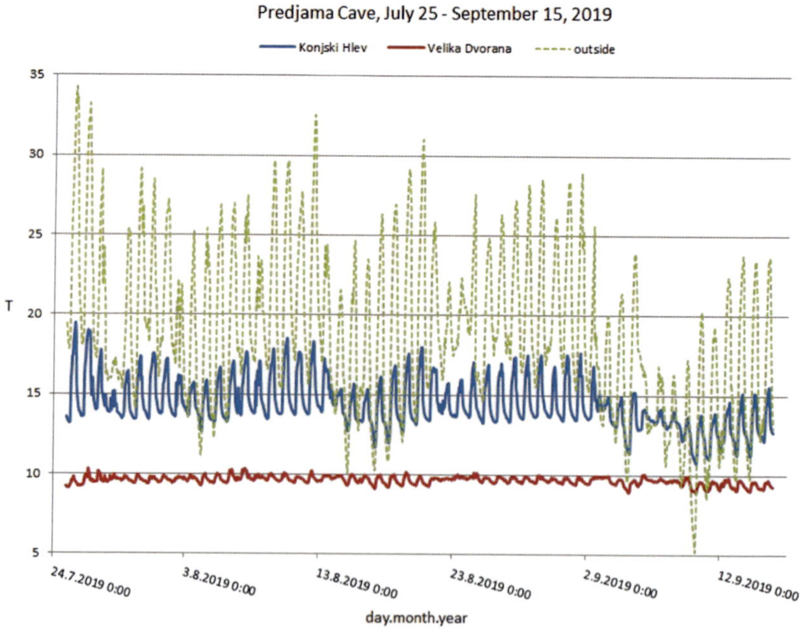

Figure 5.2.1.8 Summer hourly characteristics of air temperature in °C in Predjama Cave and outside for July 25 – September 15, 2019.

from the surface is unable to raise the air temperature at Velika Dvorana significantly (Šebela and Turk, 2014b). Conditions in Velika Dvorana are characteristic of a "climate inversion," because the cave air is equilibrated at around 9–10°C (Šebela and Turk, 2014b). Summer heating of air temperature in Velika Dvorana is an indicator of global temperature changes. If we present mean summer temperature values (June–October) for the period 2010–19 (Fig. 5.2.1.9) we can see that 2012 and 2013 had the lowest temperatures, while 2015 and 2014 had the highest temperatures. The value of 0.78°C is the greatest difference between 2013 and 2015. It can be asserted that after 2014 the outside air temperature had a stronger impact on summer cave temperature in Velika Dvorana than in the summer periods of 2010–13. This cannot be attributed to visitors, but to natural outside climate conditions.

The summer and winter ventilation regime is illustrated in Fig. 5.2.1.10. Longitudinal section AB is taken from Chapter 5.2 (Fig. 5.2.1). The situation in 2009–20 is not significantly different to the work of Anelli (1941–44). In the winter period, cold outside air enters Predjama Cave through the lower entrances and relatively warm cave air exits the cave at higher entrances (in Fiženca). At Vetrovna Luknja, the air movement direction in winter is toward Črna Dvorana. During the 2009–12 period, the measured air velocity was 1–5.5 m/s (Šebela and Turk, 2014b). The highest air movement speed recorded at Vetrovna Luknja reached 5.5 m/s on August 28, 2009 (Šebela and Turk, 2014b). However, we detected also

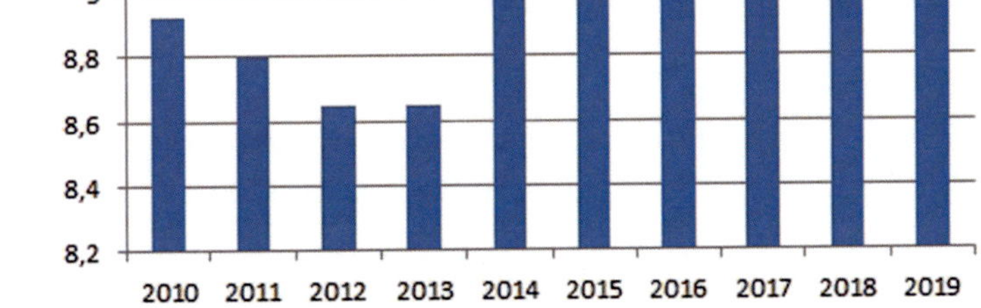

Figure 5.2.1.9 Mean air temperature in °C for June – October, 2010 – 2019 period for Velika Dvorana.

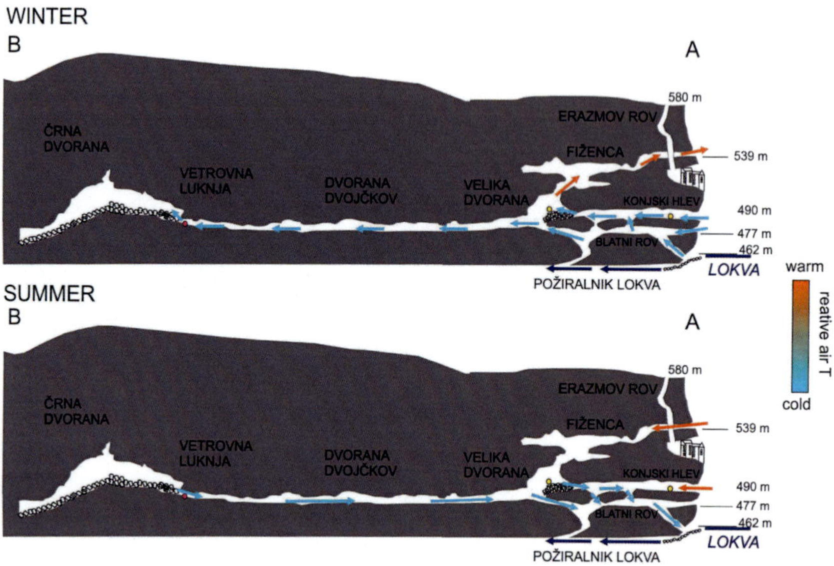

Figure 5.2.1.10 Winter and summer ventilation regimes in Predjama Cave. Arrows point the direction of the ventilation.

exceptional situations that are not in accordance with the typical summer or winter ventilation regimes, showing transition conditions (Šebela and Turk, 2014b). The reversal of the characteristic summer and winter ventilation regimes is typical not only in spring and autumn but can also occur on a daily basis (Šebela and Turk, 2014b). Such cases have already been described by Kranjc (1983).

For the summer ventilation regime, it is typical that warm outside air enters the cave through upper cave entrances (Fiženca and also Konjski Hlev). The direction of air currents at Vetrovna Luknja is the opposite to during winter, because during summer air currents flow from Črna Dvorana toward Dvorana Dvojčkov and Velika Dvorana. In summer, relatively cold cave air exits the Predjama Cave through the lower entrances (Fig. 5.2.1.10). In summer, at the Velika Dvorana monitoring site air temperature remains stable at 9–10°C because relatively warm air from Fiženca is not as dense and heavy as the relatively colder cave air and does not impact the Velika Dvorana monitoring site in summer.

CHAPTER 5.2.2

Carbon dioxide monitoring

Microclimatic monitoring in Predjama Cave does not include as many monitoring sites as at Postojna Cave. The reason is that Predjama Cave is not as heavily visited by tourists as Postojna Cave. This is the main reason for carbon dioxide measurements not being performed as extensively as in Postojna Cave. However, the first results of air temperature monitoring in Predjama Cave (Šebela and Turk, 2011b) showed that this is a complex cave system with a complex microclimatic situation, which is not limited only to the entrance (tourist) areas, but also to the rest of the cave system. Challenging situation for organization of monitoring represents interior of the Predjama Cave, which is suitable for trained speleologists.

Carbon dioxide monitoring in Predjama Cave has been organized only sporadically. Some measurements have been performed between Velika Dvorana chamber and Fiženca passage, and at Vetrovna Luknja. Here we present details on carbon dioxide levels for the period October 22–November 2, 2009 (Fig. 5.2.2.1) measured at half-hour intervals. Carbon

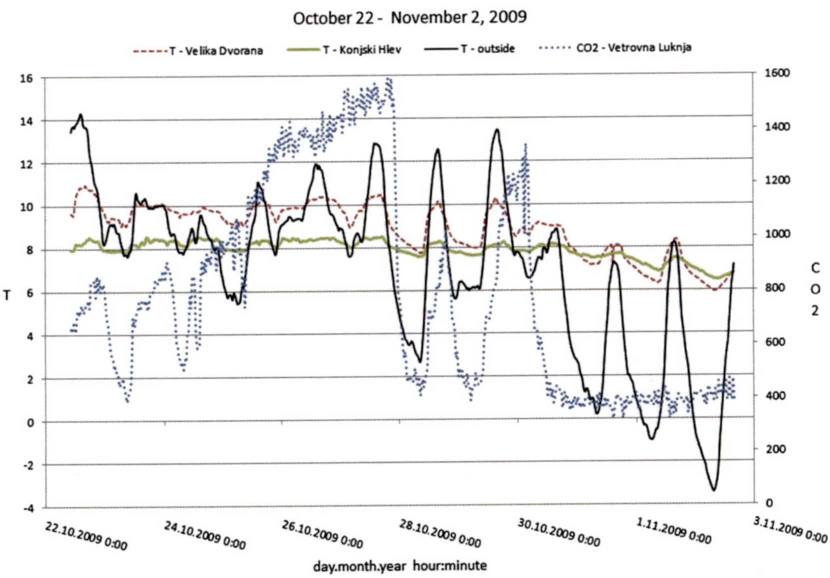

Figure 5.2.2.1 Carbon dioxide concentrations (ppm) at Vetrovna Luknja site compared with Konjski Hlev, Velika Dvorana and outside air temperatures (in °C).

Natural and Anthropogenic Impacts on Cave Climates
ISBN 978-0-12-822954-5
https://doi.org/10.1016/B978-0-12-822954-5.00024-X

dioxide concentrations were measured at Vetrovna Luknja (see Fig. 5.2.1 in Chapter 5.2, Predjama Cave), which is a vertical blowhole chimney passage, about 10 m high, 1–2 m wide (Figs. 5.2.2.2 and 5.2.2.3), and located about 600 m from the cave entrance. Originally this chimney was blocked within collapsed blocks. Historic sources (Valvasor, 1689; Nagel, 1748; Schmidl, 1854) describe strong air currents at Vetrovna Luknja causing torches be extinguished. This strong ventilation was the sign for the early explorers and speleologists that Predjama Cave must continue behind Vetrovna Luknja. Numerous old inscriptions on the cave wall in front of Vetrovna Luknja date back to the 16th (Fig. 5.2.2.4), 17th, and 18th centuries, proving that the first

Figure 5.2.2.2 Upper part to Vetrovna Luknja chimney in Predjama. *(Photo: S. Šebela.)*

Figure 5.2.2.3 Lower part of the Vetrovna Luknja vertical passage with access to Črna Dvorana collapse chamber. *(Photo: S. Šebela.)*

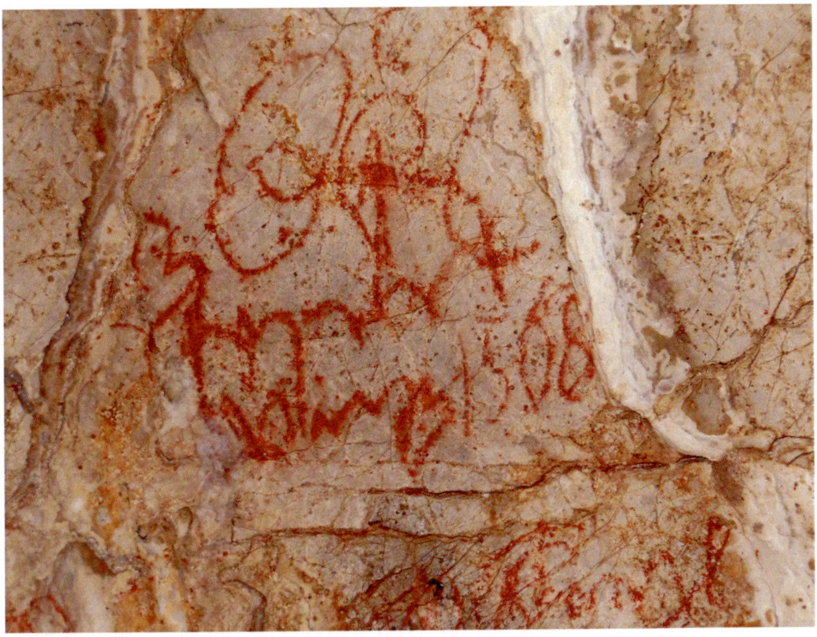

Figure 5.2.2.4 Supposably the oldest inscription (1508) in Predjama Cave on the limestone wall in front of Vetrovna Luknja. *(Photo: S. Šebela.)*

visitors stopped at Vetrovna Luknja because it was not possible to continue along the path. In 1912, the passage to Črna Dvorana was opened, collapsed blocks removed, and new passages in the system explored (Anelli, 1941–44). The newly discovered Črna Dvorana (black chamber) was so-named due to the occurrence of thin black deposits which can be found on speleothemes.

The air temperature was not measured in parallel with carbon dioxide at the Vetrovna Luknja site. Because of this, we used the temperature data set for Konjski Hlev and Velika Dvorana (Fig. 5.2.2.1) and for outside the cave. From October 22 to 27, 2009, carbon dioxide concentrations increased from 400–600 to 1600 ppm (Fig. 5.2.2.1). On October 27, 2009, at around 21:00 p.m. the carbon dioxide level started to decrease and concentrations fell to 480 ppm on October 28, 2009, at around 4:30 a.m. On October 27, 2009, the outside air temperature started to decrease at 16:00 p.m. (12.76°C), which is 5 h before the carbon dioxide level at Vetrovna Luknja started to decrease. On October 28, 2009, at 8:00 a.m., the outside air temperature reached a minimum value of 2.8°C.

The maximum outside air temperature on October 28, 2009, at 15:30 p.m., was 12.44°C, while maximum carbon dioxide concentrations of

1000 ppm were reached at 19:30 p.m. on the same day. There are 4 h difference in the daily peak between the outside air temperature and carbon dioxide level at Vetrovna Luknja.

On October 29, 2009 (Fig. 5.2.2.1), the maximum daily outside temperatures were measured at 15:30 p.m. (13.37°C), while the carbon dioxide peak was reached on October 30, 2009, at 3:30 a.m. (1340 ppm), which is 12 h after the outside cave temperature peak. From 3:30 a.m. to 14:00 p.m. on October 30, 2009, the carbon dioxide level dropped to 910 ppm. From October 30 to November 2, 2009, carbon dioxide concentrations remained stable at around 400 ppm.

The behavior of carbon dioxide levels at the Vetrovna Luknja monitoring site (Fig. 5.2.2.1) shows transitions between summer and winter ventilation regimes. Until October 27, 2009, high carbon dioxide concentrations of up to 1600 ppm were typical for summer ventilation ($T_{out} > T_{cave}$), when air currents flow from Črna Dvorana through Vetrovna Luknja toward Velika Dvorana collapsed chamber (see Fig. 5.2.1.10 in Chapter 5.2.1, Air temperature monitoring). The large drop in carbon dioxide values at Vetrovna Luknja on October 27–28, 2009, is related to an outside temperature drop of 10°C and to the transition to a winter ventilation regime ($T_{out} < T_{cave}$), when air currents flow through Vetrovna Luknja to Črna Dvorana (see Fig. 5.2.1.10 in Chapter 5.2.1, Air temperature monitoring). The daily transition between summer and winter ventilation regimes was detected on October 28–29, 2009. Meanwhile, after October 30, 2009, Predjama Cave was under a winter ventilation regime, when carbon dioxide concentrations at Vetrovna Luknja were stabilized to the outside atmospheric levels (~400 ppm) and the upward airflow mode (UAF) was presented (Faimon et al., 2012).

Pearson correlation coefficients (Table 5.2.2.1) for the period October 22–November 2, 2009, between carbon dioxide concentrations at Vetrovna Luknja and outside air temperatures give a weak (medium) correlation

Table 5.2.2.1 Pearson correlation coefficients (PCC) for carbon dioxide concentrations and air temperatures in Predjama Cave for the period October 22–November 2, 2009.

	PCC
Vetrovna Luknja CO_2–outside T	0.58
Vetrovna Luknja CO_2–Velika Dvorana T	0.67
Vetrovna Luknja CO_2–Konjski Hlev T	0.65

0.90–0.99 (very strong correlation)
0.70–0.89 (strong correlation)
0.40–0.69 (medium correlation)

(PCC = 0.58). Winter and summer ventilation regimes are strongly connected with outside air temperatures, but the PCC does not show sufficient correlation. The reason for this is the time lag between the maximum outside air temperature and the highest carbon dioxide concentrations. Vetrovna Luknja monitoring site is situated about 600 m deep inside Predjama Cave and a 5–12 h lag time between temperature and carbon dioxide peaks has been confirmed.

Predjama Cave has dynamic air ventilation because of the numerous cave entrances at different heights. Strong air currents (maximum speed 5.5 m/s, average 2 m/s) blow during winter from Vetrovna Luknja toward Črna Dvorana (Šebela and Turk, 2014b). During our visits to Predjama Cave (2009–16) occasional measurements of air current speed at Vetrovna Luknja have been accomplished with a hand-held anemometer with a precision of 0.5 m/s (Fig. 5.2.2.5). When the outside air temperature and cave air temperature are added (Velika Dvorana and Konjski Hlev) to air current speed measurements, we have additional proof of the summer and winter ventilation regimes. When $T_{out} < T_{cave}$ (blue arrows in Fig. 5.2.2.5) the air current velocity reached 5–5.5 m/s and upward airflow mode (UAF, Faimon et al., 2012) took place, as is typical for winter ventilation. During the summer ventilation regime ($T_{out} > T_{cave}$) the air current velocity was 2–5 m/s and the air flow direction was a downward airflow mode (DAF; Faimon et al., 2012).

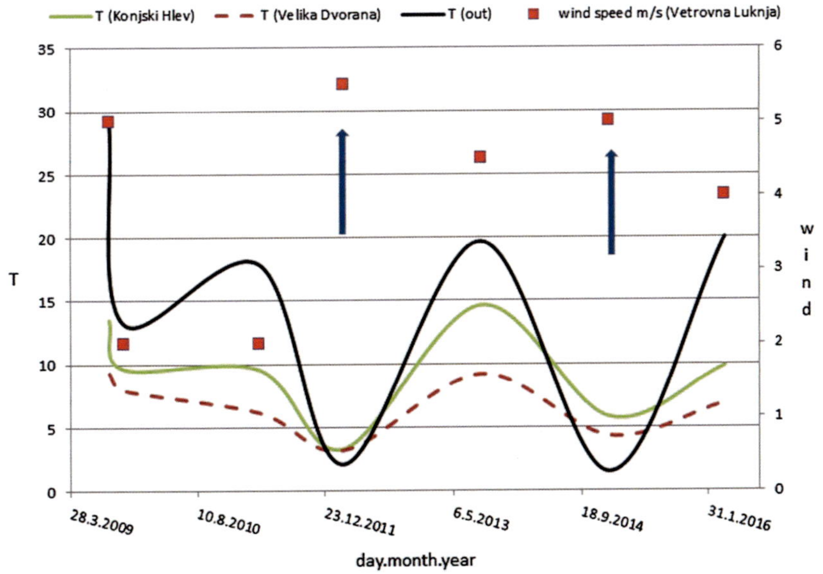

Figure 5.2.2.5 Air temperature values at Konjski Hlev, Velika Dvorana and outside (in °C) compared with velocity of air currents at Vetrovna Luknja site (m/s). Blue arrows point to winter ventilation regime periods with upward airflow mode (UAF).

Occasional carbon dioxide and air current velocity measurements at Vetrovna Luknja confirmed that summer (DAF airflow mode; Faimon et al., 2012) and winter (UAF airflow mode; Faimon et al., 2012) ventilation regimes are present deep inside the Predjama Cave.

In 1980, air temperature, air current velocity, and humidity were measured at Vetrovna Luknja (Kranjc, 1983) and in the Predjama Cave entrance part called Golobja Luknja, which is situated about 5 m behind the doors to Konjski Hlev. The mean annual air temperature at Vetrovna Luknja was 7°C and 7.2°C at Golobja Luknja in 1980. The temperature range at Vetrovna Luknja was 3.5–10°C. Fifty measurement surveys of air current speed and direction at Vetrovna Luknja were performed in 1980 (Kranjc, 1983). The strongest air currents were detected in January and December and the weakest in May and October. The mean air current speed was 4 m/s with the strongest of 9 m/s recorded on January 9, 1980 (Kranjc, 1983).

CHAPTER 5.2.3

Visitors to Predjama Cave and castle

In the case of Predjama tourist visits to the cave and also to the castle are analyzed together. Visits to the castle are not limited, while visits to the cave are limited in number. A small number of the visitors to Predjama castle equates to the whole number of Predjama Cave visitors. Tours to the cave are restricted to 50–60 visitors on a single tour because the tourists need to wear helmets with lights and there are only 60 helmets available. In Predjama Cave there is no permanent electricity, and the paths used by tourists are not paved (Fig. 5.2.3.1). At the entrance to Konjski Hlev there

Figure 5.2.3.1 Konjski Hlev entrance parts with touristic path (1 m wide) and remains of historic electric installation on the ceiling. *(Photo: S. Šebela.)*

Natural and Anthropogenic Impacts on Cave Climates
ISBN 978-0-12-822954-5
https://doi.org/10.1016/B978-0-12-822954-5.00025-1

are the remains of the historic electric installation from the end of the 19th century. Tourists in Predjama Cave follow a route that is about 700 m long, with a vertical range of 50.74 m (Šebela and Turk, 2014b). Cave tours are organized between May 1 and September 30, to protect bats hibernating during the winter months, with trips lasting about 45 min (Šebela and Turk, 2014b). During the peak tourist season there can be two to three trips to the cave daily, what means a maximum of ~200 visitors per day.

Old inscriptions on the cave walls and historic documents prove that visits to Predjama Cave have been taking place for at least 500 years. In the work *Carniola antiqua et nova*, Schönleben (1681) described the passages of Predjama Cave. Schönleben (1681) wrote that visitors entered the cave carrying lights and could hear the rush of a mighty stream. Habe (1987) described that Schönleben (1681) referred to the tale of the Argonauts, which had some connection with the stream in Predjama.

Also, Valvasor (1689) visited Predjama Cave and described Lokva stream having a special cave and that underground water emerged as Vipava springs.

In 1737, two English travelers, Richard Pecocke and Jeremiah Milles, traveled across Europe. In a letter written by Milles he describes the mountainous and bare, rarely overgrown landscape in the vicinity of the village Luek (Predjama) and the cave with one of the most romantic entrances he had ever seen (Shaw, 2008).

As already mentioned in Chapter 5.2.1, the court mathematician of Emperor Franz I, Nagel (1748), visited Predjama Cave and left significant inscriptions on a cave wall (see Fig. 5.2.1.4 in Chapter 5.2.1, Air temperature monitoring). Hacquet (1778–1789) described Predjama Cave as a cave with stalactites and mud, which ends in a narrow hole through which only water can pass.

The Windischgrätz family was at one time the owner of the castle and cave, and made the first tourist paths in Predjama Cave in the beginning of the 19th century (Habe, 1987). In 1846, when Fiženca was discovered, Weriand Windischgrätz provided ladders to connect Velika Dvorana with Fiženca (Habe, 1987).

Modern organized visits to Predjama Cave started in 1991 (Fig. 5.2.3.2) with the establishment of Slovenia's independence. The highest number of visitors to Predjama Cave, at 12,300–13,900 per year (Fig. 5.2.3.3), was registered between 2000 and 2003 (Šebela and Turk, 2014a, b). More recently there have been ~10,000 visitors per year. Predjama Cave is not a typical show cave because it remains a wild show cave with an unpaved path, ladder to the upper passages, and illumination is only with helmet lights.

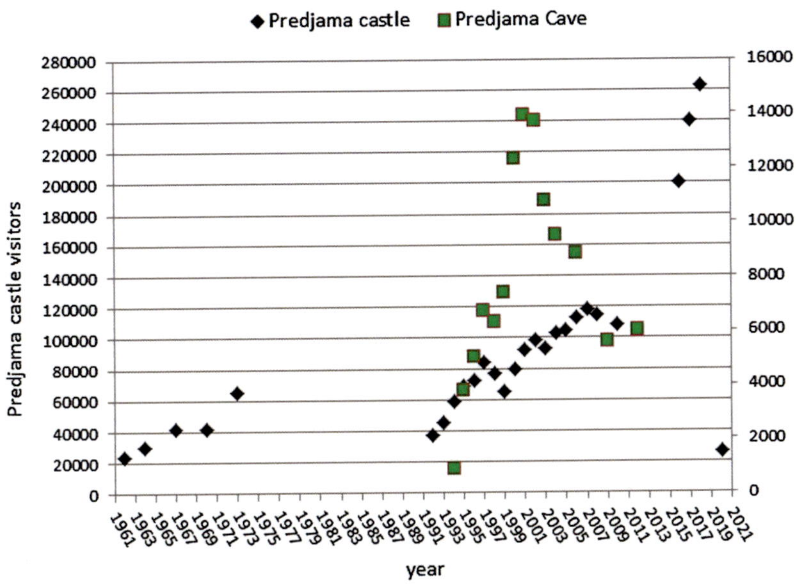

Figure 5.2.3.2 Predjama Cave and Predjama castle visitor numbers per year (1962–2020).

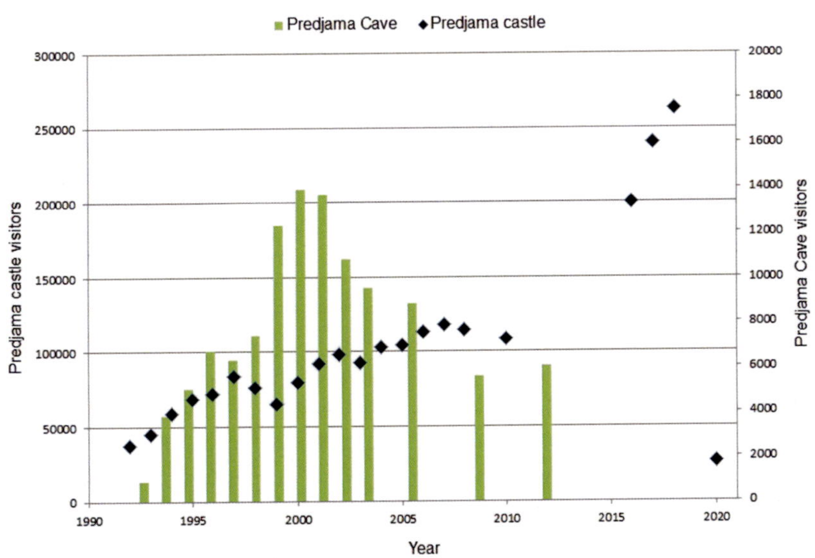

Figure 5.2.3.3 Predjama Cave and Predjama castle visitor numbers per year (1991–2020).

If the data show that the number of visitors to Predjama Cave is not increasing, this cannot be said for Predjama castle. Since the end of the Second World War to 2012 about 3 million visitors have visited Predjama castle (Šebela and Turk, 2014b).

During June—August 1973, nearly 35,000 tourists visited Predjama castle out of a total of 65,186 (Fig. 5.2.3.2) for the whole year (Habe, 1987). In 1992, the number of visitors dropped to 37,600 due to the 1991 war in the former-Yugoslavia. After 1992, visitors to Predjama castle returned and tourism increased. In August 2010 there were 26,633 visitors to Predjama castle. In a single day on August 19, 2017, there were 2892 visitors to the castle and on August 15, 2018 it had 3522 visitors (http://www.noviceznotranjske.net/park-postojnska-jama-s-14-000-obiski-v-enem-dnevu-znova-podrl-rekord/). Since 2010 there has been a very significant increase such that in 2016 200,000 visitors was reached, with 263,000 visitors in 2018. In 2020, with the impact of the COVID-19 pandemic a huge decrease in visitor numbers is expected (−90%).

It was published in 2014 that Predjama castle occupied fifth or sixth place among the most-visited Slovenian tourist destinations (Šebela and Turk, 2014b). We can see that in recent years (Fig. 5.2.3.4) visits to Predjama castle have increased even further, and that this castle is the fourth-placed tourist destination in Slovenia after Ljubljana castle, Postojna Cave, and Bled castle. It is clear that some of the Postojna Cave visitors also visit Predjama castle.

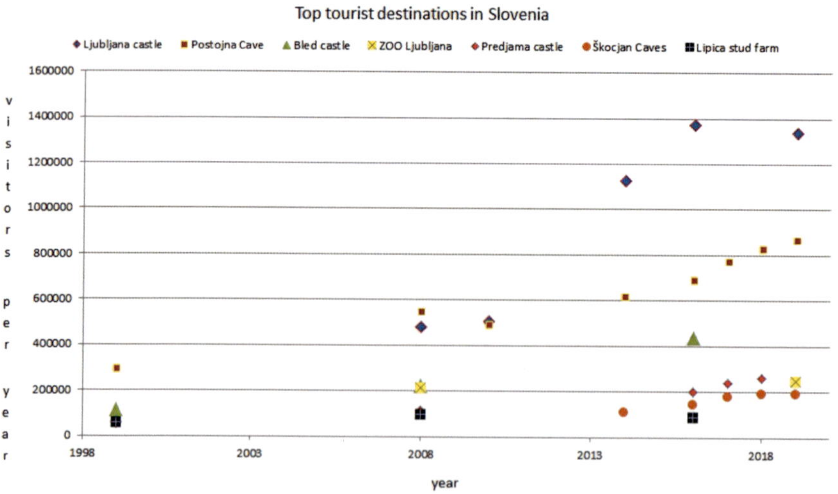

Figure 5.2.3.4 Top tourist destinations in Slovenia 1999—2019.

Before 2008 Postojna Cave was the top tourist destination in Slovenia (Fig. 5.2.3.4), and Predjama castle was fifth or sixth. Since 2011 Ljubljana castle has taken first place, with Postojna Cave the second most visited destination in Slovenia.

CHAPTER 5.2.4

Natural impacts on the Predjama Cave microclimate

Due to the very low number of visitors to Predjama Cave (see Chapter 5.2.3, Visitors in Predjama Cave) we can discuss only the natural impacts on the cave's microclimate, because the anthropogenic impact is negligible. What are the natural drivers that influence the Predjama Cave microclimate? The first factor is morphology of cave passages and cave entrances at different elevations. The second factor is the presence of summer [downward airflow mode (DAF) according to Faimon et al., 2012] and winter [upward airflow mode (UAF); Faimon et al., 2012] ventilation regimes, which show an exchange between outside and cave air masses. The most important driver of the Predjama Cave microclimate is the temperature difference between the cave air and outside air.

The mean annual air temperatures at the two monitoring sites in Predjama Cave and outside the cave (Fig. 5.2.4.1) show similar behavior of temperature curves from 2010 to 2019. After 2014, all three temperature curves have very

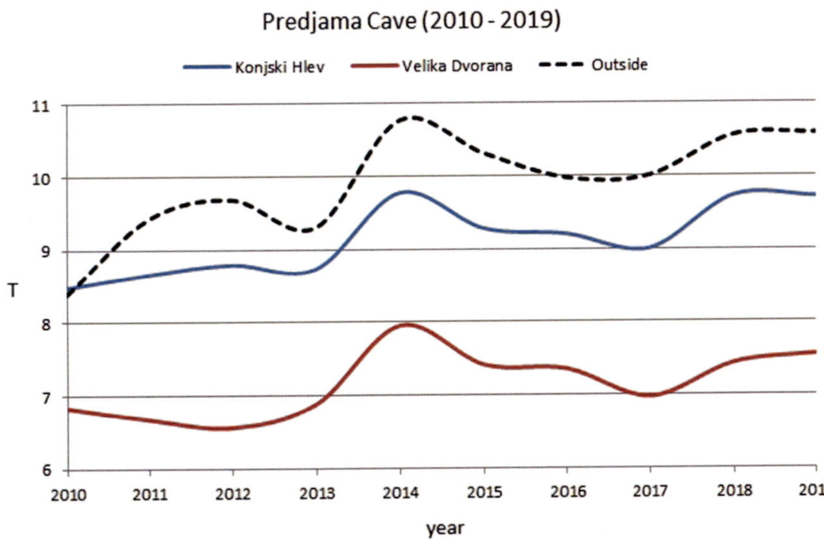

Figure 5.2.4.1 Mean annual air temperature (°C) in Predjama Cave and outside for the period 2010–2019.

Natural and Anthropogenic Impacts on Cave Climates
ISBN 978-0-12-822954-5
https://doi.org/10.1016/B978-0-12-822954-5.00026-3

similar patterns of ups and downs. The Pearson correlation coefficient (PCC) between Velika Dvorana and outside mean annual air temperatures for the period 2014–19 was 0.84, while it was 0.77 for 2010–19 (Table 5.2.4.1). For the first part of the monitoring period 2010–14, the Pearson correlation coefficient (PCC) between Velika Dvorana and outside mean annual air temperature was 0.71. This demonstrates that the outside climate conditions had a smaller impact on the Velika Dvorana monitoring location in the period 2010–2014 than was the situation in 2014–19. The Konjski Hlev site always shows a very strong correlation with the outside conditions. During 2010–19, the PCC between Konjski Hlev and outside was 0.93 (Table 5.2.4.1), and it was 0.92 for the period 2010–14 and 0.94 for the period 2014–19. The correlation between two Predjama Cave monitoring sites is very strong (PCC = 0.9) for the period 2010–19 and for the period 2010–14 (PCC = 0.92). However, for the years 2014–19 the correlation is weaker (PCC = 0.82).

Since 2011, the mean annual outside air temperature has been consistently higher than it at Konjski Hlev and Velika Dvorana (Fig. 5.2.4.2). In 2010, the mean air temperature outside the cave was slightly lower than in Konjski Hlev, but after 2011 the mean annual air temperature outside the cave was consistently higher (up to 1.03°C in 2015) than in Konjski Hlev (Fig. 5.2.4.2). The temperature difference between outside the cave and the Velika Dvorana site was even higher than in Konjski Hlev. The difference between mean annual air temperature outside the cave and Velika Dvorana was up to 3.12°C (in 2012 and 2018). After 2012, the temperature difference did not increase and showed only smaller changes.

Calculation of the mean air temperature values for the period 2010 to 2019 shows that long-term air temperature in the Konjski Hlev monitoring site was 0.76°C lower than outside the cave (Table 5.2.4.2). In Velika Dvorana the difference (−2.73°C) was even higher. This negative difference shows that the Predjama Cave passages are not overheated. Even the Konjski Hlev monitoring site, which is situated close to the cave entrance does not exhibit increased long-term air temperature as is found in the environment outside the cave.

Table 5.2.4.1 Pearson correlation coefficients (PCC) for mean annual air temperatures in Predjama Cave for 2010–2019.

	PCC
Konjski Hlev–outside	0.93
Konjski Hlev–Velika Dvorana	0.90
Velika Dvorana–outside	0.77

0.90–0.99 (very strong correlation)
0.70–0.89 (strong correlation)

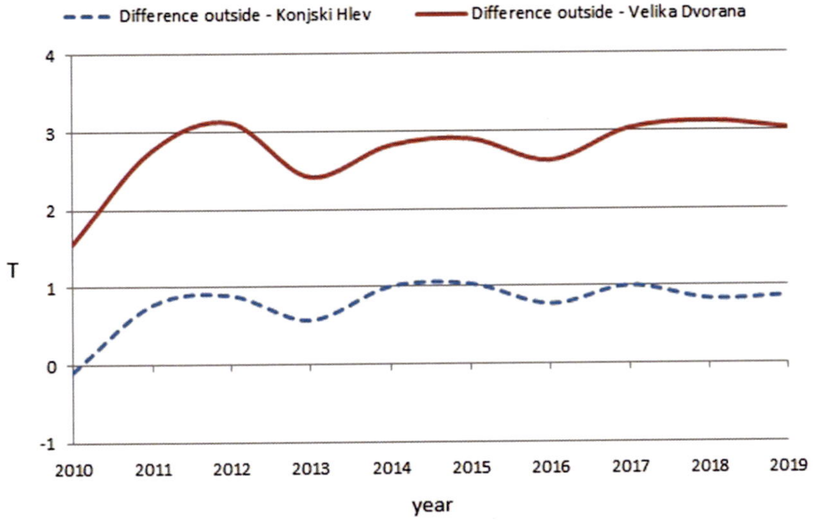

Figure 5.2.4.2 Difference between mean annual air temperature (°C) outside the cave and in Predjama Cave for the period 2010–2019.

Table 5.2.4.2 Difference between outside and cave air temperature in °C for mean temperature values 2010–2019.

	Konjski Hlev	Velika Dvorana	Outside
Mean *T* (2010–2019)	9.13	7.16	9.89
Difference outside – cave	−0.76	−2.73	–

We can see that mean annual air temperatures in Predjama Cave are increasing (Fig. 5.2.4.1), and also that the outside cave mean annual temperature values are increasing. The temperature increase for Konjski Hlev site is more significant than it is for Velika Dvorana. The percentage temperature increase has been calculated according to the mean annual air temperature values for 2019 subtracted by the mean long-term (2010–19) temperature values. The results show that the temperature increase for the outside cave atmosphere in 2019 was 6.97% higher than the long-term (2010–19) mean value. Similarly, the calculated air temperature for Konjski Hlev in 2019 was 6.35% higher than the long-term (2010–19) mean value. The Velika Dvorana monitoring site shows the lowest air temperature increase. In 2019, a temperature rise of 5.44% was calculated for Velika Dvorana for the long-term (2010–19) mean value. Temperature increases for two monitoring sites in Predjama Cave are attributed to natural conditions and not to anthropogenic impacts inside the cave. The increase in outside air temperature causes an increase in air temperatures in

Predjama Cave also. Here, we do not look to find evidence for anthropogenic and/or natural causes of the outside air temperature increase in the period 2010–19; our goal remains study of the cave microclimate.

Air temperature monitoring in Predjama Cave is a good reference site to detect any temperature increase in the cave due to an increase in the outside cave temperature, because the number of visitors is negligible and the anthropogenic impact does not need to be considered. Increased values obtained for Predjama Cave can be compared to Postojna Cave, where increased temperatures include natural and anthropogenic impacts.

In the past, the mean annual air temperatures at Vetrovna Luknja measured in 1980 (Kranjc, 1983) were 7°C and 7.2°C in Golobja Luknja (5 m deep inside the cave and behind the doors to Konjski Hlev, respectively). The temperature range in 1980 at Vetrovna Luknja was 3.5–10°C and at Golobja Luknja it was −11.5 to 17.7°C (Kranjc, 1983). It can be seen that the long-term (2010–19) mean air temperature at Velika Dvorana was 7.16°C (Table 5.2.4.3), which is comparable with the mean annual temperature at Vetrovna Luknja in 1980. However, we do not have the temperature data for 2010–19 for the Vetrovna Luknja site. On the other hand, the Golobja Luknja site can be paralleled with Konjski Hlev because both sites are only about 20 m apart. The mean annual air temperature at Golobja Luknja in 1980 was 7.2°C (Kranjc, 1983), while at Konjski Hlev it was 9.13°C (2010–19). If we have a 6.35% temperature increase in Konjski Hlev (Table 5.2.4.3) between 2010 and 2019, do we get a 26% temperature increase in Golobja Luknja between 1980 and 2019? It is clear that the position of Golobja Luknja very close to the cave entrance must mean that it is greatly affected by outside climate changes.

In addition to, air temperature monitoring, air pressure is measured at hourly intervals at the Predjama Cave locations (Fig. 5.2.4.3). In the first half of 2019 the synchronization between outside and cave air pressure locations was good, with higher air pressure values at Konjski Hlev than in Velika Dvorana. After July 2019, the instrument in Konjski Hlev provided inconsistent measurements and in the second half of 2019 only instruments in Velika Dvorana and the outside monitoring locations performed air pressure monitoring.

Table 5.2.4.3 Temperature increase in Predjama Cave and outside the cave since 2010.

	Konjski Hlev	Velika Dvorana	Outside
Mean T [2010–2019 (°C)]	9.13	7.16	9.89
Mean T [2019 (°C)]	9.71	7.55	10.58
T increase (%)*	+6.35	+5.44	+6.97

* T increase = mean T (2019) − mean T (2010–2019)

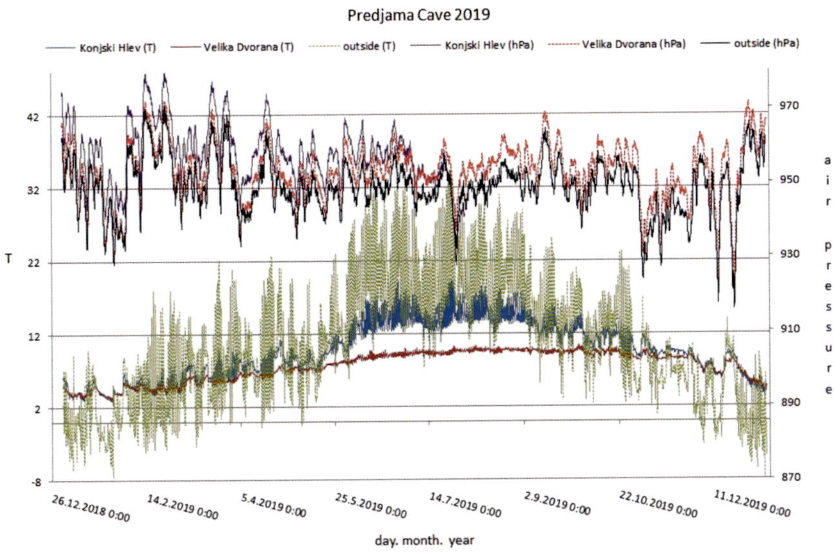

Figure 5.2.4.3 Air pressure (hPa) and air temperature (°C) in 2019 at Predjama Cave monitoring sites and outside the cave.

The PCC for hourly air pressure measurements in Predjama Cave for 2019 (Table 5.2.4.4) give a very strong correlation between Velika Dvorana and outside (PCC = 0.97). The correlation between Konjski Hlev air pressure measurements and the outside location presents a strong correlation (PCC = 0.89). A very good correlation between cave sites and outside is another factor showing the strong influence of the outside climate on Predjama Cave. To detect some possible air pressure differences between a cave and the outside environment, air pressure measurements should be more frequent than hourly intervals (Šebela and Turk, 2011b).

Table 5.2.4.4 Pearson correlation coefficients (PCC) for hourly air pressure measurements in Predjama Cave for 2019.

	PCC		Period
Velika Dvorana—outside	0.97		2019
Konjski Hlev—Velika Dvorana	0.89		January 1—August 2, 2020
Konjski Hlev—outside	0.92		January 1—August 2, 2020

0.90—0.99 (very strong correlation)
0.70—0.89 (strong correlation)

CHAPTER 5.2.4.1

Difference between day and night air temperatures in Predjama Cave

The correlation between midnight and midday air temperatures for Konjski Hlev and Velika Dvorana is presented in Figs. 5.2.4.1.1 and 5.2.4.1.2. The position of the temperature monitoring site at Konjski Hlev, which is placed close to the cave entrance, is reflected as the difference between midday and midnight temperature measurements, showing a significant impact of the outside climate on the cave air temperature. The greatest difference between midday and midnight reached up to 6°C in 1 day (Fig. 5.2.4.1.1). The temperature curves for Velika Dvorana monitoring site (Fig. 5.2.4.1.2) do not show as high a difference between midday and midnight temperatures as is the case for the Konjski Hlev site. Velika Dvorana has very stable climate conditions in the relationship between

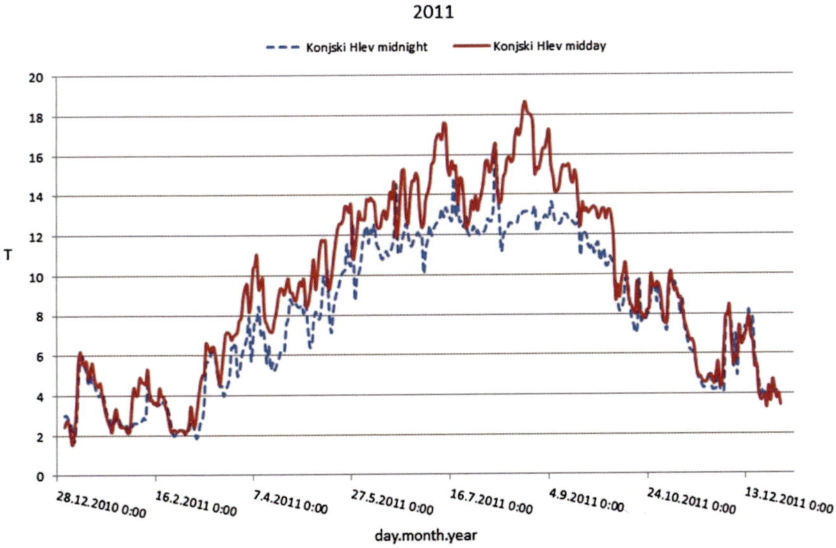

Figure 5.2.4.1.1 Relation between midnight and midday air temperatures in 2011 at Konjski Hlev in °C.

Natural and Anthropogenic Impacts on Cave Climates
ISBN 978-0-12-822954-5
https://doi.org/10.1016/B978-0-12-822954-5.00027-5

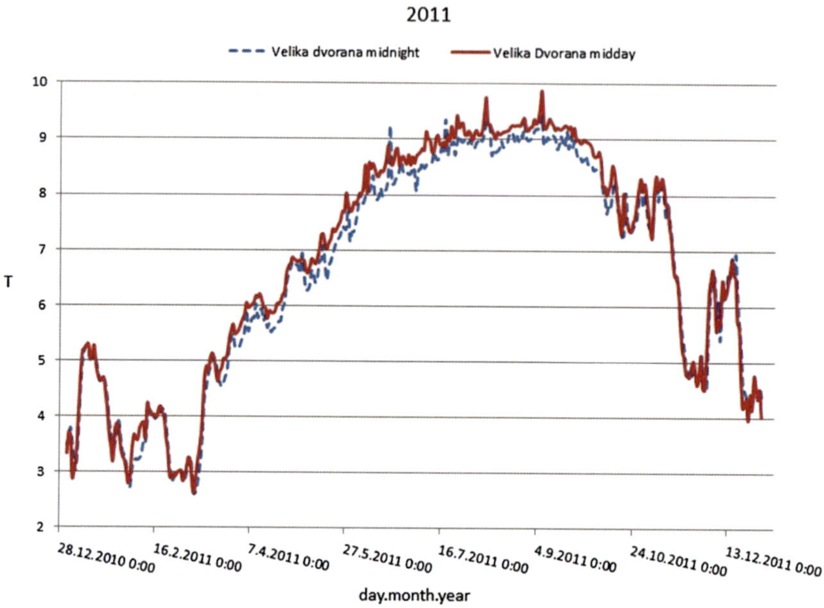

Figure 5.2.4.1.2 Relation between midnight and midday air temperatures in 2011 at Velika Dvorana in °C.

midday and midnight temperatures. The Pearson correlation coefficient (PCC) for Konjski Hlev is 0.96, which is a very strong correlation between the midday and midnight air temperatures (Table 5.2.4.1.1). The PCC for Velika Dvorana is 0.99 and gives an almost total positive linear correlation for the year 2011, meaning that between midnight and midday the air temperature measurements showed almost no difference.

In Chapter 5.1.7.1 (Difference between day and night air temperatures in Postojna Cave) we presented examples for two different years, 2011 and 2018, and for Predjama Cave we also give information for 2018. Comparing the Konjski Hlev situations in 2011 and 2018 it can be seen that in the 2011 air temperature was between 2 and 18.5°C (Fig. 5.2.4.1.1) while due to the colder winter it was 0–18.5°C in 2018 (Fig. 5.2.4.1.3). The midnight air temperature in 2011 at Konjski Hlev was generally

Table 5.2.4.1.1 Pearson correlation coefficients (PCC) between midday and midnight air temperatures in 2011 and 2018 at Konjski Hlev and Velika Dvorana.

		PCC (2011)	PCC (2018)
Konjski Hlev	Midnight–midday	0.96	0.96
Velika Dvorana	Midnight–midday	0.99	0.99

0.90–0.99 (very strong correlation)
0.70–0.89 (strong correlation)
0.40–0.69 (medium correlation)

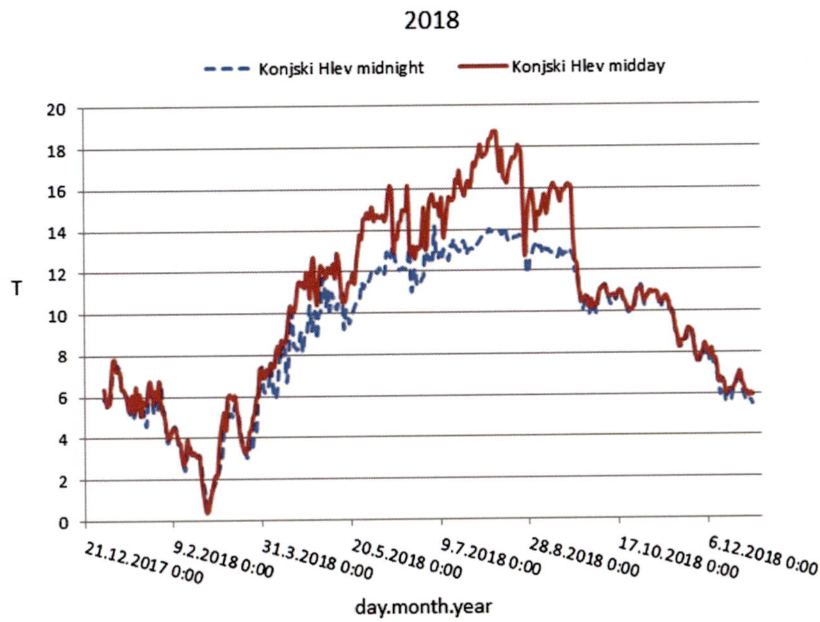

Figure 5.2.4.1.3 Relation between midnight and midday air temperatures in 2018 at Konjski Hlev in °C.

~13°C, with a maximum peak of up to 16°C. In 2018, the summer midnight air temperature at Konjski Hlev was ~13–14°C, meaning that it was at least 0.5°C higher in 2018 than in 2011. The Velika Dvorana monitoring site summer midnight air temperatures in 2011 were around 9°C, with some peaks of up to 9.3°C (Fig. 5.2.4.1.2). In 2018, midnight air temperatures at the same location generally were ~9.5°C with peaks of up to 9.8°C (Fig. 5.2.4.1.4). Midday summer temperatures at Velika Dvorana were stabilized at ~9.3°C, with peaks of up to 9.8°C in 2011. In 2018, midday summer temperature values were just below 10°C, with some peaks slightly above 10°C. Summer midnight air temperatures at Velika Dvorana chamber were about 0.5°C higher in 2018 than in 2011. Midday temperatures in summer 2018 at Velika Dvorana were 0.6–0.7°C higher than in 2011. This is clearly seen in Fig. 5.2.4.1.5. Although the winter February temperatures in 2018 were lower than in 2011, summer and autumn values were characteristically higher in 2018 than in 2011 at Velika Dvorana.

Pearson correlation coefficients (PCC) between midday and midnight air temperatures at Konjski Hlev and Velika Dvorana were equal in 2011

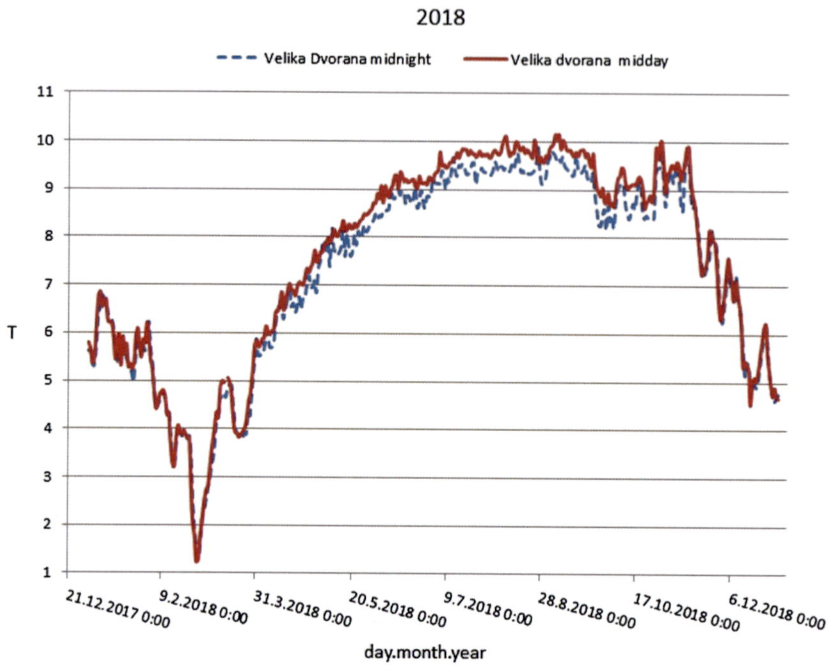

Figure 5.2.4.1.4 Relation between midnight and midday air temperatures in 2018 at Velika Dvorana in °C.

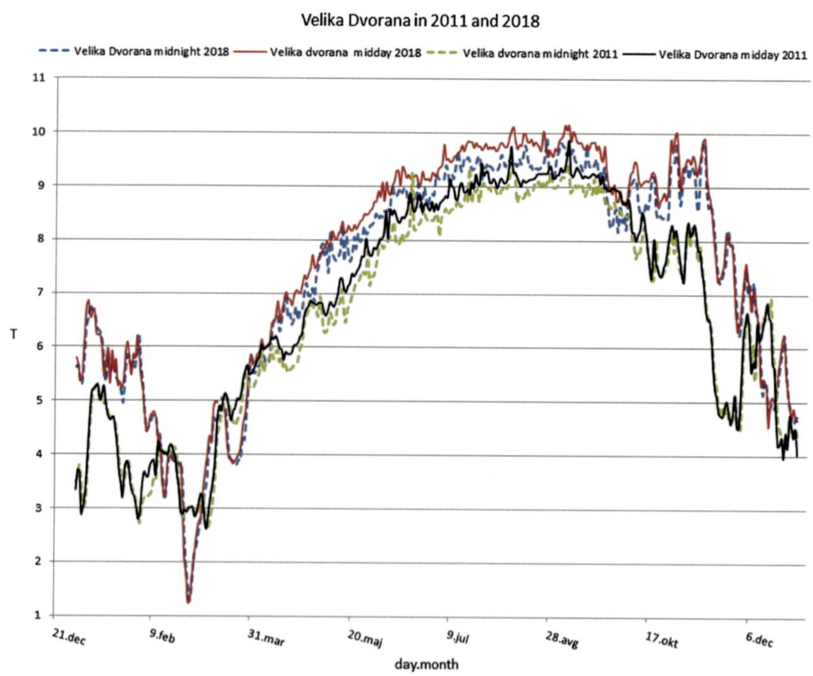

Figure 5.2.4.1.5 Comparison between midnight and midday air temperatures in Velika Dvorana in 2011 and in 2018 in °C.

and 2018 (Table 5.2.4.1.1). In Chapter 5.1.7.1 (Difference between day and night air temperatures in Postojna Cave) all calculated Postojna Cave sites (Postojna 1, Postojna 2, Postojna 3, Koncertna Dvorana, Sepolcro, and Pulpito) showed a very strong correlation between day and night temperatures, which is also the case in Predjama Cave.

In Chapter 5.1.7.1 (Difference between air temperatures in Postojna Cave) the difference between midday and midnight values was compared between 2011 and 2018. Midnight temperature values at the Velika Gora site (Postojna 1) were 0.47°C–0.66°C higher in 2018 than in 2011, and midday temperatures were 0.52°C–0.68°C higher in 2018 than in 2011. This is very similar to the case in Predjama Cave. It must not be forgotten that the increase in the air temperature from 2011 to 2018 at Postojna Cave is attributed to an increase in the outside temperature, and to some extent also to the increased number of visitors, while in Predjama Cave the anthropogenic impact cannot be included and the air temperature increase at Predjama Cave from 2011 to 2018 is purely a consequence of an outside temperature increase.

CHAPTER 5.2.4.2

Extreme natural events in Predjama Cave

The Lokva stream that runs on Eocene flysch on the northwest side of the Pivka valley sinks at 462 m above sea level into the Predjama Cave system. This is generally a small watercourse, which can become dangerous with extremely high precipitation. Habe (1970) described Lokva water flow rates of 5 L/s during droughts and 200 L/s during high water periods.

There is historic evidence of high floods in front of Predjama Cave (Šebela, 2012). It appears that the oldest document referring to floods is from 1826 (Habe, 1966). Lokva supposedly reached a level of 495 m above sea level, as the anonymous witness described water reaching above the wall in Golobja Luknja (Habe, 1966). The same catastrophic event in 1826 caused the highest water level of the Reka River in Škocjan Caves (Habe, 1966; Kranjc, 1998), which shows that this was a regional catastrophic event. It was published in 2012 (Šebela) that the level of 495 m for the 1826 floods (Habe, 1966) was probably too high for Predjama Cave, because the wall at Golobja Luknja is situated at about 490–491 m (Šebela, 2012). Schmidl (1854) described a high water level in 1851 at Konjski Hlev due to the flooding of Lokva. During this event wooden parts of the mill on Lokva were washed into the Blatni Rov passage (Schmidl, 1854).

On September 2, 1965, heavy precipitation (more m^3/s) caused the formation of a lake in front of the Predjama Cave system (Habe, 1970). The water was 2 m lower than the bridge in front of Konjski Hlev and reached 488 m above sea level (Habe, 1966). The same event caused very high registered floods at Škocjan Caves on the same day, where the water height was increasing at the rate of 5 m/h (Habe, 1966). The direct distance between Predjama and Škocjan Caves is ~20 km.

The most recent floods date to September 19, 2010 (Fig. 5.2.4.2.1), when the water in Konjski Hlev reached 489.60 m, and 489.1 m in Velika Dvorana (Šebela, 2011, 2012). This was also a regional event as floods covered a large part of southern, southwestern, and southeastern Slovenia. For karstologists and speleologists, it was for first time that the blocking of the Predjama Cave entrance passages with high water had been witnessed.

Natural and Anthropogenic Impacts on Cave Climates
ISBN 978-0-12-822954-5
https://doi.org/10.1016/B978-0-12-822954-5.00028-7

Figure 5.2.4.2.1 Muddy remains on vegetation on September 20, 2010 in front of the Predjama Cave after the retreat of high waters. *(Photo: S. Šebela).*

The air temperature datalogger in Konjski Hlev was flooded (Šebela, 2012) but it continued to measure the water temperature instead of air temperature. It was calculated that the floods in Konjski Hlev probably lasted 40 h and 30 min or 44 h in the wider observation window (Šebela, 2012). The water reached the datalogger in Konjski Hlev on September 18, 2010, at between 3:50 a.m. and 7:20 a.m., and reached the lower cave passages on September 19, 2010, at 23:50 p.m. (Šebela, 2012).

The water level of the September 19, 2010 floods at Predjama was about 1 m lower than at the Konjski Hlev passage and stabilized at 489.6 m (Šebela, 2012), which is 1.6 m higher than the floods on September 2, 1965 (Habe, 1970). The highest floods were in 1826 (Habe 1966), reaching around 495 m or more probably 490–491 m above sea level (Šebela, 2012).

The lower levels of Velika Dvorana and Stara Jama all the way to Vetrovna Luknja (see Fig. 5.2.1 in Chapter 5.2, Predjama Cave) were under water. This was not clear but very muddy water (Fig. 5.2.4.2.1) mostly due to the flysch material in it. After the withdrawal of the water, it was seen that the "Slovenski gadje 1882" charcoal inscription on the limestone wall below Vetrovna Luknja was damaged (Figs. 5.2.4.2.2 and 5.2.4.2.3) due to abrasion by the high waters that reached about 485 m at Vetrovna Luknja site (Šebela, 2012). Regarding the studied historic records, the September 2010 floods were the highest in Predjama at least since 1882 (Šebela, 2012) and most probably were the second highest since the well-documented floods in 1826, which were described by Habe (1966).

After the retreat of the water in the late night of September 19, 2010, or in the morning of September 20, 2010, a collapse in cave sediments (see Fig. 5.2.1 in Chapter 5.2, Predjama Cave), 5 × 4 × 0.3 m in dimensions, occurred (Šebela, 2011) in Konjski Hlev (Fig. 5.2.4.2.4). In Slovene caves, sudden collapses of cave floor are very rare. In this case we witnessed in "real time" a sudden collapse due to the withdrawal of flooded water.

About 12 m below the collapse there are passages of Zmajeva Luknja at 477 m above sea level (Habe, 1970). Between Konjski Hlev and Zmajeva Luknja there must be widened subvertical fissures or even shafts filled with cave sediments and not only hard rock limestone. Through one of these

Figure 5.2.4.2.2 »Slovenski gadje 1882« charcoal inscription below Vetrovna Luknja in August 2009 before it was damaged by high waters in September 2010. *(Photo: S. Šebela.)*

Figure 5.2.4.2.3 »Slovenski gadje 1882« charcoal inscription below Vetrovna Luknja was damaged after retreat of high waters in September 2010. *(Photo: S. Šebela.)*

Figure 5.2.4.2.4 Collapse in Konjski Hlev after retreat of high waters in September 2010. *(Photo: S. Šebela.)*

forms flooded water rose from Zmajeva Luknja to upper Konjski Hlev and formed a small lake in Konjski Hlev. After the retreat of this water a sudden collapse occurred in the area of the small lake (Šebela, 2011). Over the following 10 years the collapsed area did not become larger or deeper. The tourist path that leads close to the collapse has not been destroyed by recent collapse activity.

Beside floods, freezing is widely present at Predjama Cave's entrance areas during cold winters. With winter ventilation, cold outside air penetrates deep into the cave system (see Fig. 5.2.1 in Chapter 5.2, Predjama Cave), as far as Dvorana Dvojčkov (Habe, 1970) and Vetrovna Luknja (Šebela et al., 2017).

Dripstone fall observations over a 1-year period (1980) were made at the entrance to Predjama Cave (Kranjc, 1983). Mean dripstone fall from the cave wall was 0.2 $g/m^2/day$ (Kranjc, 1983). Dripstone fall was the most intensive during late winter and the lowest during summer and early autumn. The main reason for dripstone fall is the freeze–thaw effect, when the temperature moves above and below 0°C (Kranjc, 1983).

CHAPTER 6

Sustainable use of show caves

More than 20% of the Earth's crust is characterized by epigean and/or hypogean karst phenomena (Ford and Williams, 2007). Karst and caves are extremely valuable natural resources important for tourism, economy, science, and education. There are more than 1200 show caves open worldwide (Garofano and Govoni, 2012) and over 250 million visitors per year buy a ticket to visit show caves (Cigna and Forti, 2013). There are an estimated 200 million people directly or indirectly earning their income from show caves (Cigna and Forti, 2013). On the other hand, caves are very fragile environments, which may be easily destroyed when they are transformed into a tourist attraction (Cigna and Forti, 2013). Thus, it is very important to follow the recommendations for sustainable use of caves for tourism (Cigna and Forti, 2013). The best Environmental Impact Assessment (EIA) starts with a pre-operational phase of the show cave (Cigna and Buri, 2000), but some show caves have already been opened for 200 years (Šebela et al., 2015a) or longer and EIA at the pre-operational phase is therefore not possible.

In the case of Postojna and Predjama Caves, current expert control was established in 2009 based on a short-term program (2009–13) for the sustainable use of natural assets for tourism (Culver et al., 2012). The most important tasks of this short-term program (2009–13) were: active involvement of the cave custodian in environmental issues of the show cave, biological and climatic monitoring, water monitoring, lampenflora monitoring, regularly informing the Regional Unit of the Institute of the Republic of Slovenia for Natural Conservation regarding the implementation of the concession contract and the state of the caves, implementing improvement programs for the elimination of impacts caused by the use of the natural assets, calling attention to external factors that can harm the caves, monitoring and supervision of large-scale interventions and performances in caves, providing evidence of research, providing an inventory of interventions in caves, and providing an inventory of the natural and cultural heritage of caves (Culver et al., 2012). Within the long-term program (2009–28) for the use of natural assets, the role of cave custodian also includes natural protection and precautions based on the

Natural and Anthropogenic Impacts on Cave Climates
ISBN 978-0-12-822954-5
https://doi.org/10.1016/B978-0-12-822954-5.00029-9

results of 5-year monitoring of the cave climate and cave fauna (Culver et al., 2012).

The cave custodian or cave guardian for Postojna Cave is an independent public research institution (ZRC SAZU Karst Research Institute) with rich national and international experience and a multidisciplinary approach to karstology (Blatnik et al., 2020).

Biological monitoring is organized to determine the current state of the Postojna and Predjama Caves ecosystems and to make an inventory of animals (Figs. 6.1—6.3) that live in caves and for which caves represent an important habitat for their life cycle. It is important to monitor different species of cave animals as indicators of general changes in the whole cave systems (Culver et al., 2012). Monitoring of fauna in percolated waters is included within biological monitoring. Certain species can be used as indicator species for estimation of impacts on the environment (Culver et al., 2012). Systematic studies have already been accomplished, and most indicator species have already been defined in Postojna Cave, which is also known as the cradle of speleobiology (Culver and Pipan, 2009).

The intention of climatic monitoring in show caves is to monitor any short-term and long-term changes in caves due to tourist use of the cave systems, and especially the impact of tourists on the cave climate (Culver et al., 2012; Šebela et al., 2015a). To address this, in 2009 we began systematic climate monitoring of basic parameters (air temperature, air pressure, humidity, CO_2 in the air, and air circulation) for selected segments of

Figure 6.1 Bat in Predjama Cave. *(Photo: S. Šebela.)*

Figure 6.2 *Meta menardi* spider in Postojna Cave. *(Photo: S. Šebela.)*

Figure 6.3 *Proteus anguinus* in Postojna cave. *(Photo: S. Šebela.)*

the Postojna and Predjama Caves. In first years, studies were oriented to heavily visited cave passages. Later, microclimatic monitoring was extended to non-touristic parts of the cave systems to quantify the natural and/or human causes of microclimatic changes.

After the last group leaves the temperature in Postojna Cave drops toward the value recorded before the thermal disturbance caused by visitors (Šebela et al., 2015a). The temperature recovery is gradual and can be interrupted by a high number of visitors entering the cave the following day (Šebela et al., 2015a). A minor cumulative effect of cave air temperature increase due to visitors of 0.14°C was observed at the beginning of January 2014 (Šebela et al., 2015a). Generally the cumulative effect (of air temperature and carbon dioxide) is relatively small in Postojna Cave, due to the natural ventilation toward the exterior (Šebela et al., 2015a).

Artificial illumination in show caves is responsible for the growth of autotrophic photosynthetic organisms called lampenflora that develop in close and remote proximity to lights (Mulec and Kosi, 2009). These phototrophic organisms cause the degradation of cave deposits, which is a particular problem in caves with prehistoric art and speleothems (Mulec and Kosi, 2009). Key factors that allow lampenflora to grow are light and moisture (Mulec and Kosi, 2009). In the lampenflora community, various aerophytic algae, as well as some mosses and ferns, dominate (Mulec and Kosi, 2009).

The simplest solution to preventing lampenflora growth in a cave would be complete removal of existing phototrophic communities, cessation of the illumination, and abolition of tourist visits, which is unacceptable to the cave management (Mulec and Kosi, 2009). The reduction of light intensity and duration, as well as the use of light-emitting diodes (LEDs) to control lampenflora growth, are suggested in show caves (Mulec and Kosi, 2009). For suppressing lampenflora growth, the use of hydrogen peroxide (H_2O_2) instead of an aggressive 5% aqueous solution of NaOCl has been suggested (Faimon et al., 2003). This method has been successfully applied in Postojna Cave and Škocjan Caves in Slovenia (Mulec, 2014). Bearing in mind that permanent installation of electric lighting in Postojna Cave has been present since 1884 (Šebela, 2019) and that in recent years LEDs have been installed in Lepe Jame passage, treatment to reduce lampenflora is a very important subject in the management of Postojna Cave.

Since 1872, a railway has been in operation for tourists in Postojna Cave (Čuk, 2008; Šebela, 2019). In 1924, the man-powered railway was replaced by the first locomotive powered one (Čuk, 2008; Šebela, 2019). The cave train has always been an important tourist attraction in Postojna Cave. However, operation of the cave train produces dust particles, which remain in the cave. This was described by Gams (1974) and Muri et al. (2013). During a period of high tourist numbers, the rise in CO_2 levels due to anthropogenic influences was less significant than the increase in dust particles in the cave atmosphere caused by the scattering of fresh sand over the railway tracks (Gams, 1974). In 2018 new locomotives with improved passenger car wheel were introduced to Postojna Cave (Figs. 6.4 and 6.5), which significantly lowered the input of dust and non-cave material to the cave environment. Before the renovation of the railway in Postojna Cave the pre-2018 railway was an important anthropogenic source of particulate matter in the cave (Muri et al., 2013; Šebela et al., 2015a).

Figure 6.4 Cave train to Postojna Cave used before 2018. *(Photo: S. Šebela.)*

Figure 6.5 Cave train to Postojna Cave used after 2018. *(Photo S. Šebela.)*

The growth of speleothems in a show cave might be at risk from effects related to increased visits. This risk is higher in caves with poor ventilation and where either the calcium ion concentration of the drip waters is low or where there have been significant changes in surface land use, which decreases the drip-water calcium concentrations and is more likely to cause corrosion (Baker and Genty, 1998). In Postojna Cave most speleothem damage is due to cave visitors in the past who broke speleothems to sell or keep as souvenirs.

With systematic biological and climatic monitoring in the most visited show caves in Slovenia (Postojna Cave and Škocjan Caves) the increasing number of visitors became an issue as the carrying capacity arose. In 2019, in the pre-COVID-19 pandemic period, Škocjan Caves hosted 200,000 and Postojna Cave 870,000 visitors. The visitor capacity can be defined as the maximum number of visitors acceptable in a time unit under defined conditions which does not imply a permanent modification of the relevant parameters (Cigna and Buri, 2000). The maximum numbers of visitors to Postojna Cave per day and per year are linked to logistics, because tourists are brought to the interior of the cave by train, and the train passenger-carrying capacity is limited (Šebela et al., 2015a). The COVID-19 pandemic situation has dramatically lowered (~85%) the number of visitors to Postojna Cave in 2020. The tourist-carrying capacity for Postojna Cave has not yet been actualized, but a maximum of 7700 visitors per day (August 15, 2014) and 870,000 visitors per year (2019) are in the range of limited numbers.

Karst caves in Slovenia are natural assets that belong to the State. Postojna Cave is managed by a private company and local municipality, while Škocjan Caves are a regional park under the authority of the Ministry of the Environment and Spatial Planning. Management of the two principal show caves in Slovenia has two perspectives (private and park service), where both must work toward sustainable use of the caves and mitigation of the side effects of long-term tourism. Successful management of the show cave must combine productive communication and interaction between researchers, cave owners, cave managers, the public, and the State government (Šebela et al., 2015a).

Regarding the results of hourly air temperature monitoring in Postojna (see Chapter 5.1.7, Anthropogenic and natural impacts on Postojna Cave microclimate) and Predjama Caves (see Chapter 5.2.4, Natural impacts on Predjama Cave microclimate) for the period 2010 to 2019, the mean annual air temperatures have been calculated using cave and outside monitoring sites and, compared with the number of visitors to Postojna Cave per year, the number of visitors to Predjama Cave is negligible (<10,000 per year). The air temperature increase was calculated as a percentage to see the relationship between Postojna Cave, Predjama Cave, and outside conditions (Fig. 6.6). The absolute difference between mean air temperature in 2019 and mean air temperature in 2010 shows a higher temperature increase in Predjama Cave than Postojna Cave. There was a 76% increase in visitor numbers to Postojna Cave between 2010 and 2019. Also, the outside air temperature shows a high increase (26%) in 2019 compared with 2010.

Similar relations are seen by comparing mean air temperatures for 2019 with mean air temperatures for the entire period of 2010–19 (and for 2011–19 for Koncertna Dvorana). The increase in air temperature at the Predjama Cave sites reflects an increase in the outside air temperature and is not connected with the presence of visitors. In contrast, we would expect a higher air temperature increase at Postojna Cave due to the very high increase in tourist numbers, but Fig. 6.6 does not show this. All Postojna

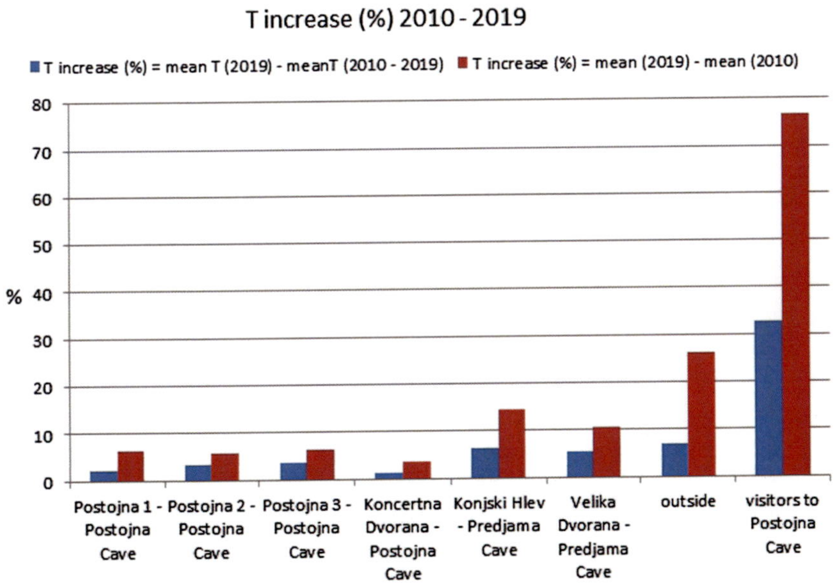

Figure 6.6 Air temperature and visitors' increase (%) for the period 2010 to 2019 (Koncertna Dvorana for 2011 to 2019).

Cave temperature monitoring sites presented in Fig. 6.6 are situated 2–3 km deep inside the cave, while the Predjama Cave sites are closer to the cave entrance (50–200 m), which may mean that the inner Postojna Cave passages, situated in deep karst massive, have been more isolated and resistant to temperature entry for the calculated period of 2010–2019 than the Predjama Cave sites.

However, there is another perspective, which must not be forgotten. The mean annual cave air temperature is supposed to be similar to the mean annual outside air temperature (Smithson, 1991). If we compare Postojna Cave with Predjama Cave, we can see that Postojna Cave is overheated in comparison with the outside conditions (see Fig. 5.7.1.1 in Chapter 5.7.1, Anthropogenic and natural impacts on Postojna Cave microclimate) and Predjama Cave is hypothermic in comparison with the outside conditions (Fig. 6.7). This graph gives a more realistic view of the impact of visitors to Postojna and Predjama Caves. Even if Predjama Cave has been heating faster than Postojna Cave since 2010 (Fig. 6.6) due to the increase in outside air temperature it is still colder than Postojna Cave (Fig. 6.7), which shows the accumulated heat deep inside the Postojna Cave system (probably

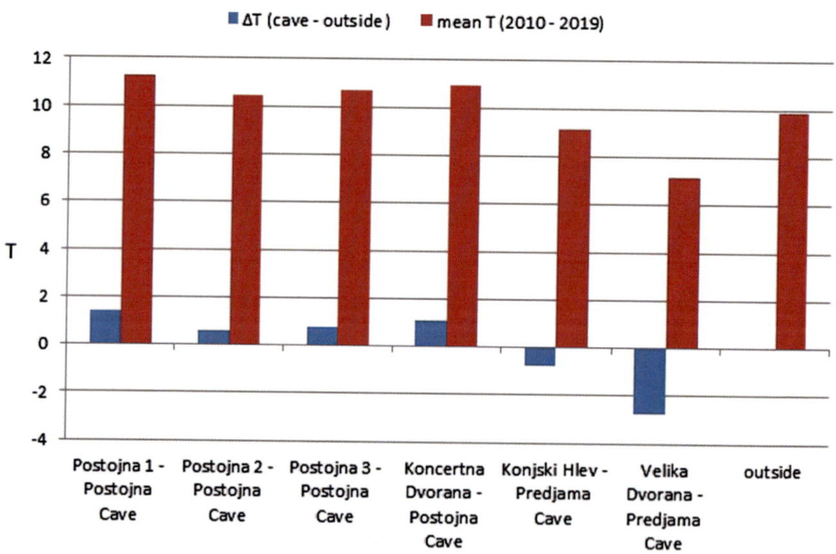

Figure 6.7 Mean air temperature in °C for the period 2010–19 and difference between the mean cave air temperature and outside air temperature for the period 2010–19 (Koncertna Dvorana for 2011–19).

existing before 2010), is partly due to an increase in the outside air temperature and also due to the impact of visitors, including heat from people and lights. In this sense we can compare the heavily visited show cave (Postojna Cave) with a less visited show cave (Predjama Cave) to see the anthropogenic and natural impacts on the cave microclimate.

CHAPTER 7

Conclusions

Microclimatic and meteorological conditions in Postojna and Predjama show caves (Slovenia) are discussed based on hourly air temperature data and carbon dioxide concentrations. For selected periods, additional measurements of air pressure and air circulation were performed in order to explain specific meteorological conditions. Microclimatic monitoring at both show caves was established in 2009 as part of expert control of the use of caves for tourism. In 2009, three air temperature monitoring sites (Postojna 1—Velika Gora, Postojna 2—Lepe Jame, and Postojna 3—Lepe Jame, see Fig. 4.1 in Chapter 4, Methods) were managed in the most visited part of Postojna Cave. In 2020, 14 active air temperature monitoring locations carried out hourly measurements in Postojna Cave. Monitoring has been extended to a location with almost no tourists (as Otoška Jama) and another with low tourist numbers (as Črna and Pivka Jama) in the Postojna Cave system in order to see the differences between the natural and anthropogenic impacts on cave climates.

In Predjama Cave two air temperature monitoring sites have been continuously operating since 2009 (see Fig. 5.2.2 in Chapter 5.2 Predjama Cave). Tourist visits to Predjama Cave (~10,000 in 2019) are very low compared to Postojna Cave (870,000 in 2019). Comparing the microclimate situation between these two caves for the entire monitoring period of 2009–2020 provided significant facts about the conditions in heavily and poorly visited show caves over the last 10 years (see Chapter 6, Sustainable use of show caves). To see the influence of outside climate conditions on the cave microclimate, an air temperature monitoring site was established in 2009 in the forest above Postojna Cave.

The main conclusions of the research include:

- Air temperature values at Postojna Cave monitoring locations suggest that cave air circulation depends on the temperature difference between the cave and outside conditions. In Postojna and Predjama Caves there is convective air circulation. In winter the cave blows air out at higher cave entrances and takes in air at the lower cave entrances. The opposite occurs in summer. This was described with three ventilation modes, based on intensive exchange between cave air and outside air: UAF

Natural and Anthropogenic Impacts on Cave Climates
ISBN 978-0-12-822954-5
https://doi.org/10.1016/B978-0-12-822954-5.00030-5

mode (upward airflow), DAF mode (downward airflow), and transitional mode (Faimon et al., 2012, Lang et al., 2017a, b). When $\Delta T < 0$ ($\Delta T = T_{out} - T_{cave}$), upward air flow (UAF mode) occurs and when $\Delta T > 0$, downward air flow (DAF mode) takes place (Faimon et al., 2012, Lang et al., 2017a, b).

- Extensive exchange between outside air and cave air in Postojna Cave, that is transferred deep inside the cave's inner parts, is additionally enabled by the artificially opened main cave entrance that has been in place since 1866. The temperature difference between the cave and outside air is the cause of summer ($T_{out} > T_{cave}$) and winter ventilation ($T_{out} < T_{cave}$) at the main cave entrance. In the case of very cold winters, the blocked entrance prevents the growth of ice stalactites and stalagmites in the Postojna Cave entrance area to some extent, but cannot completely prevent intrusion of cold outside air into the cave, because there are other lower entrances and outside air can use opened fissures and openings to enter the cave.
- Carbon dioxide concentrations in Biospeleološka Postaja (Postojna Cave) reached 4100 ppm in August 2015. This was because of the lack of natural ventilation in a small cave passage closed with metal doors and an increased number of visitors. In the sense of sustainable use of the cave for tourism, the activated mitigation arrangement (opening of metal doors in the middle of the passage and forced ventilation with the rest of the cave) lowered the CO_2 values, which reached up to 2500 ppm in August 2018. Parallel with a decrease in CO_2 values, the mean annual air temperatures were lowered by 0.49°C at the Biospeleološka 1 site and 0.57°C at the Biospeleološka 2 site in the period 2016–19.
- The Pulpito and Sepolcro monitoring sites were re-established at the same place as the historic air temperature monitoring was set (Crestani and Anelli, 1939). The Pulpito site represents the transition area between the cave entrance and inner parts because it is closer to the main Postojna Cave entrance, while the Sepolcro site is part of the inner cave. Historical cave air temperature data at Sepolcro for 1933–37 showed a temperature range of 7.5–8.4°C, which is 1.74–2.64°C lower than measurements taken from 2015 to 2017, with an average of 10.14°C (Pipan et al., 2019). This temperature increase is mostly attributed to an increase in the outside air temperature, whatever the cause, the heat added to the cave by visitors and lights cannot be completely excluded.

- The Postojna 1 air temperature monitoring site is situated 560 m above sea level on top of limestone collapsed blocks of Velika Gora collapsed chamber. The annual air temperature range at Velika Gora is 0.58°C, which is the lowest oscillation of all the monitoring sites in Postojna Cave. The mean annual air temperature at Velika Gora site shows a 1-month time-lag between the highest air temperature summer peak and the outside peak. As an annual range, Velika Gora has ~0.5°C higher air temperature than the Postojna 3 site (Lepe Jame) and 0.5–1.2°C higher than Postojna 2 (Lepe Jame). Velika Gora acts as a "warm air trap," accumulating warmer air even with the intrusion of cold outside air during winter. Historic air temperature measurements at the bottom of Velika Gora in 1852 (Schmidl, 1854) and 1933 (Crestani and Anelli, 1939) were 8.5–8.58°C, what compared to more recent measurements is ~2°C lower.
- Postojna 3 site in Lepe Jame was established close to the tourist path in the most visited passage of Postojna Cave (~1500 m far from the main cave entrance) in order to measure the anthropogenic contribution to cave air temperature and carbon dioxide concentrations. Increased visitor numbers in the last days of December 2018 (~3000 per day) increased the carbon dioxide concentrations from 1000 ppm to 1970 ppm at 17:00 p.m. [see Fig. 5.1.3.3.9 in Chapter 5.1.3.3, Lepe Jame (Postojna 3 site)]. During the same period the daily temperature increase was 0.3°C with a 6-day cumulative effect of ~0.2°C due to the anthropogenic impact. The Postojna 3 site exhibits a daily switch from winter (during nights) to summer (during days) ventilation regimes, which was determined during the spring and autumn periods.
- At the Postojna 3 site a remarkable event was detected on February 2012 (Šebela et al., 2015a). Due to a strong northern (bora) wind outside the cave and winter ventilation, the air temperature at the Postojna 3 site decreased by 0.5°C in a wide time window (February 1–12, 2012). This was the most significant and rapid drop in air temperature at this monitoring location between 2009 and 20. Extreme natural conditions can have an even greater impact on cave air temperature than anthropogenic causes due to increased visitor numbers.
- The Postojna 2 site represents a side small chamber in Lepe Jame passage, which was artificially enlarged in the past due to strong ventilation behind the chamber. This is a particularly important location because it shows the highest air temperature values in winter and the lowest in

summer. In a speleological sense this site is placed at the crossing between the Lepe Jame tourist passage and probable unknown vertical and horizontal passages behind the Postojna 2 monitoring site, resulting in microclimatic monitoring strongly supporting the idea of the existence of unknown passages behind the Postojna 2 site, similar to the case described by Liñán et al. (2020). The upward airflow mode (UAF) (Faimon et al., 2012; Lang et al., 2017a, b) during winter is pronounced at the Postojna 2 monitoring site, while summer conditions with a lack of ventilation show a "cold air-trap" situation exchanging no air with the surrounding atmosphere. The Postojna 2 monitoring site exhibits very high concentrations of carbon dioxide (up to 4000 ppm in winter 2019–20), which cannot be attributed only to an anthropogenic source of carbon dioxide due to visitors passing through Lepe Jame, but must be strongly related to a natural origin, mostly from unknown cave passages behind this site.

- The position of Koncertna Dvorana, one of the largest collapsed chambers within the Postojna Cave, is subjected to strong intrusion of cold winter air circulation and also to warmer summer ventilation. It appear that this site is placed within the main cave air circulation currents. An air temperature peak in Koncertna Dvorana occurs 2–4 weeks after an outside air temperature peak.
- Črna Jama is the coldest part of the entire Postojna Cave system. The mean annual air temperature at the Črna Jama 2 site was 5.76°C in 2019. The cave is visited by smaller groups during warmer months and is classed as a poorly visited show cave. Črna Jama exhibits winter and summer temperature modes. During warm periods ($T_{cave} < T_{out}$) Črna Jama acts as a cold air-trap, exchanging no air with the outside atmosphere. In winter, dense and cold air sinks into the cave, cooling the interior rock surface and the cave air due to unidirectional advection of cold air into the cave. Pearson correlation coefficients for 2019 showed that Črna Jama 2 and 3 sites both show characteristics of cold air-trap caves, with a medium correlation with the outside climate. The entrance to Črna Jama is developed at the bottom of a collapsed doline and faces its northern side at 533 m above sea level. The morphology of the cave entrance, arrangement of the Črna Jama passages, and closed artificial doors with the tourist part of the Postojna Cave and Pivka Jama are enabling Črna Jama to remain a cold air-trap cave.

- The air temperature measured at Pivka Jama (Črna Jama 4 site) shows a strong correlation with the active water passage of the underground River Pivka. In the period 2015–2019, the mean air temperature was very similar to outside cave conditions and 3.45°C higher than at the Črna Jama 3 site and 5.12°C higher than the Črna Jama 2 site.
- Otoška Jama receives only occasional alternative visits and is treated as a cave with almost no tourist visitors. Air temperature monitoring started on December 4, 2017 at two monitoring sites. It was found that the Otoška Jama 2 site is the warmest monitoring site in the whole of the Postojna Cave system as it has a +1.95°C higher mean annual air temperature than the outside mean annual air temperature. This is described as natural microclimate conditions, especially with the fact that a lot of heat is transmitted to the cave air from the underground River Pivka which causes the characteristics of a warm-air trap cave.
- Only the entrance part of Predjama Cave is arranged for tourist visits during warmer months. Summer and winter ventilation regimes have been determined in Predjama Cave already in the work of Anelli (1941–44). A 10-year air temperature data set (2009–20) shows that temperature values in Konjski Hlev are similar to the outside cave conditions. At Velika Dvorana chamber the summer temperature values are stabilized and do not exceed 9–10°C. Both monitoring sites exhibit lower mean annual temperatures than the outside air temperature. The increase in mean annual air temperature from 2011 to 2019 in the Konjski Hlev and Velika Dvorana monitoring sites is related to the outside cave climate conditions because the number of visitors was negligible.
- Being a show cave for more than 200 years, anthropogenic impacts on the Postojna Cave microclimate due to visitors are detected in changes in air temperature and carbon dioxide levels. These are mostly short-term impacts because after visitors leave the cave, both parameters return to their previsit values. On the other hand, the most overheated cave site is the almost tourist-free Otoška Jama 2 site with a mean temperature 12.51°C (2018–19).
- The air temperature increase was calculated as a percentage to illustrate the relationship between Postojna Cave, Predjama Cave, and outside conditions (see Fig. 6.6 in Chapter 6). The absolute difference between mean air temperature in 2019 and mean air temperature in 2010 shows a higher temperature increase in Predjama Cave (+14.5% for Konjski

Hlev and +10.5% for Velika Dvorana) than in Postojna Cave (+6.4% for Velika Gora, +5.7% for Postojna 2, +6.4% for Postojna 3 and + 3.6% for Koncertna Dvorana). There was a 76% increase in the number of visitors to Postojna Cave between 2010 and 2019. Also, the outside air temperature showed a significant increase (+26%) in 2019 compared with 2010. The increase in air temperature at the Predjama Cave sites reflects an increase in the outside air temperature and is not connected with the presence of visitors. In contrast, we would expect a greater air temperature increase at Postojna Cave due to the very high increase in the number of tourists, however Fig. 6.6 (Chapter 6, Sustainable use of show caves) does not show this. All Postojna Cave temperature monitoring sites presented in Fig. 6.6 are situated 2–3 km deep inside the cave, while the Predjama Cave sites are closer to the cave entrance (50–200 m), which may mean that the inner Postojna Cave passages, situated in deep karst massive, have been more isolated and resistant to temperature entry for the calculated period (2010–2019) than the Predjama Cave sites.

- The fact that in temperate climates mean annual cave air temperature reflects mean the outside temperature (Smithson, 1991) gives a more clear idea of the Postojna and Predjama Cave microclimate. If we compare mean annual cave air temperature with mean annual outside air temperature (see Fig. 6.7 in Chapter 6, Sustainable use of show caves) we can see that the most visited parts of Postojna Cave are overheated compared with the outside conditions (see Fig. 5.1.7.1 in Chapter 5.1.7, Anthropogenic and natural impacts on Postojna Cave microclimate) and Predjama Cave is hypothermic compared with the outside conditions (see Fig. 6.7 in Chapter 6, Sustainable use of show caves). Even if Predjama Cave has been heated more than Postojna Cave since 2010 (see Fig. 6.6 in Chapter 6, Sustainable use of show caves) due to the increase in outside air temperature, it remains colder than Postojna Cave (see Fig. 6.7 in Chapter 6, Sustainable use of show caves), which illustrates the accumulated heat deep inside the Postojna Cave system (probably existing before 2010), partly due to the increase in outside air temperature and also due to the impact of visitors, including heat from people and lights. In this sense we can compare the heavily visited show cave (Postojna Cave) with the less visited show cave (Predjama Cave) to see the anthropogenic and natural impacts on cave microclimates.

- This study shows the importance of mutual cooperation between cave managers, ministries responsible for the environment, and cave guardians (institutions with strong karstology knowledge who suggest guidelines for sustainable use of caves for tourism) for the implementation of the best measures for the sustainable use of these caves for tourism.

Appendix

Stanka Šebela
2021

Slovenia is a small European country with and area of 20,200 km^2 and two million inhabitants. It has been independent since 1991. Forty-three percent of the country is represented by karst areas (Fig. 1) with 13,660 registered caves.

This monograph primarily addresses microclimate research in two show caves in Slovenia (Postojna and Predjama Caves, Fig. 1) because they are the caves with the longest air temperature data sets, as air temperature monitoring started at hourly intervals in 2009. Additionally, some other show caves [Škocjan Caves, Županova Jama (Fig. 2) and Kostanjeviška Jama (Fig. 3)] are discussed to a lesser extent.

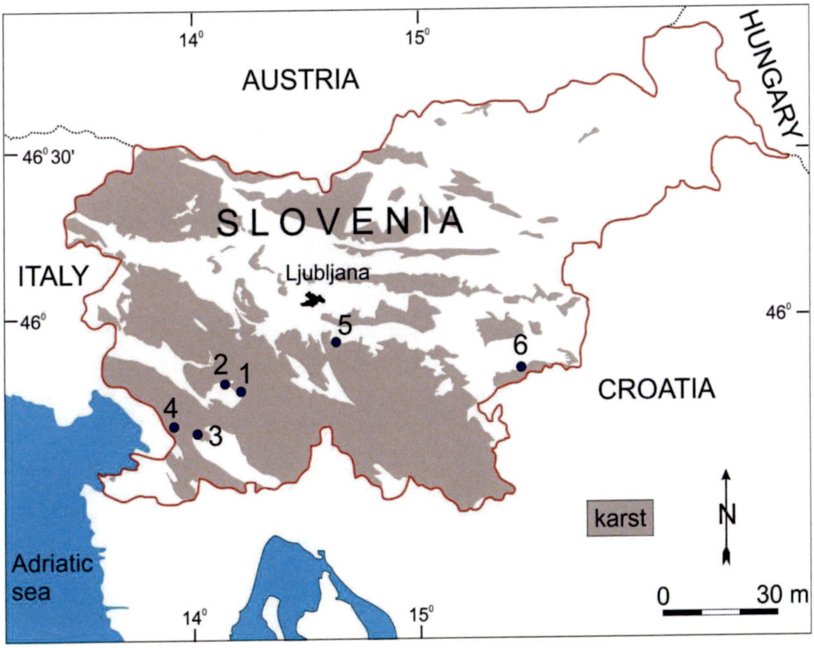

Figure 1 Karst areas in Slovenia with positions of the studied show caves. 1, Postojna Cave; 2, Predjama Cave; 3, Škocjan Caves; 4, Vilenica Cave; 5, Županova Jama Cave; 6, Kostanjeviška Jama Cave.

Figure 2 Županova Jama (710 m long and 70 m deep), number 5 in Fig. 1. *(Photo: S. Bonzanni (with permission).)*

This research work has deepened the analysis of the microclimate data obtained from field measurements in the period 2009–20, and is especially oriented toward the reciprocity between natural and anthropogenic impacts on cave climates in show caves in Slovenia. The aim of the research is for examples from microclimate studies from Slovene show caves to be used as examples for similar show caves worldwide, to strengthen international scientific cooperation, and provoke new scientific challenges.

Vilenica (Fig. 1) is the oldest show cave in Europe. In 1633 it was opened for guided tours. Modern visits have been organized since 1962.

Figure 3 Kostanjeviška Jama (1871 m long and 47 m deep), number 6 in Fig. 1. *(Photo: S. Šebela.)*

Vilenica (Fig. 4) is richly decorated with speleothems and hosts numerous shield- or palette-shaped speleothems. Occasional meteorological measurements have been performed in the cave but systematic continuous measurements are not yet available.

The karst caves in Slovenia are natural assets that belong to the State. Postojna Cave (Fig. 5) is the second longest cave in Slovenia and is managed

Figure 4 Vilenica Cave (841 m long and 190 m deep), number 4 in Fig. 1. *(Photo: S. Šebela.)*

Figure 5 Postojna Cave (24,120 m long and 115 m deep), number 1 on Fig. 1. *(Photo: S. Šebela.)*

by a private company and local municipality, while Škocjan Caves are under the remit of the Regional Park in the Ministry of the Environment and Spatial planning. Management of the two principal show caves in Slovenia has two perspectives (private and park service), where both must work toward the sustainable use of the caves and mitigation of the side effects of long-term tourism.

Škocjan Caves (Fig. 6) are the only karst caves in Slovenia that have been listed as a UNESCO's World Heritage Site since 1986. All show caves

Figure 6 Škocjan Caves (6474 m long and 168 m deep) with the underground River Reka and illuminated tourist path, number 3 in Fig. 1. *(Photo: S. Šebela.)*

need a balance between protection of natural worth and increased tourism. This can only be provided by good knowledge of the natural characteristics of the cave obtained from karst research studies, climatic and biological monitoring, and specifying the permissible level of anthropogenic impact.

Predjama Cave (number 2 in Fig. 1) has 16,755 m of known passages (168 m deep) and is the third longest cave in Slovenia. The cave is known for extensive cave passages over a number of levels, archeological remains, bat hibernation, numerous old inscriptions on the cave walls, and the history related to Predjama castle (Fig. 7). Only the entrance parts of the cave are used for tourist visits.

The purpose of research described in this monograph is to present the complexity of microclimatic research in Postojna and Predjama Caves, especially with regard to cave air temperatures and carbon dioxide concentrations. Microclimatic monitoring in show caves is organized as part of environmental protection monitoring, which started in 2009 with expert control aimed at the sustainable use of the Postojna (Fig. 8) and Predjama Caves for tourism. The goal was to see the relationship between natural and anthropogenic impacts on climates of show caves. The fact is that both causes interact, and it is not easy to distinguish between them. In this sense, the microclimate of the heavily visited Postojna Cave was compared with the less visited Predjama Cave, where the anthropogenic impact due to visitors was eliminated.

Figure 7 Predjama castle. *(Photo: S. Šebela.)*

Figure 8 Postojna Cave. *(Photo: S. Šebela.)*

The absolute difference between mean air temperature in 2019 and mean air temperature in 2010 (Fig. 9) shows a higher temperature increase in Predjama Cave (+14.5% for Konjski Hlev and +10.5% for Velika Dvorana) than in Postojna Cave (+6.4% for Velika Gora, +5.7% for

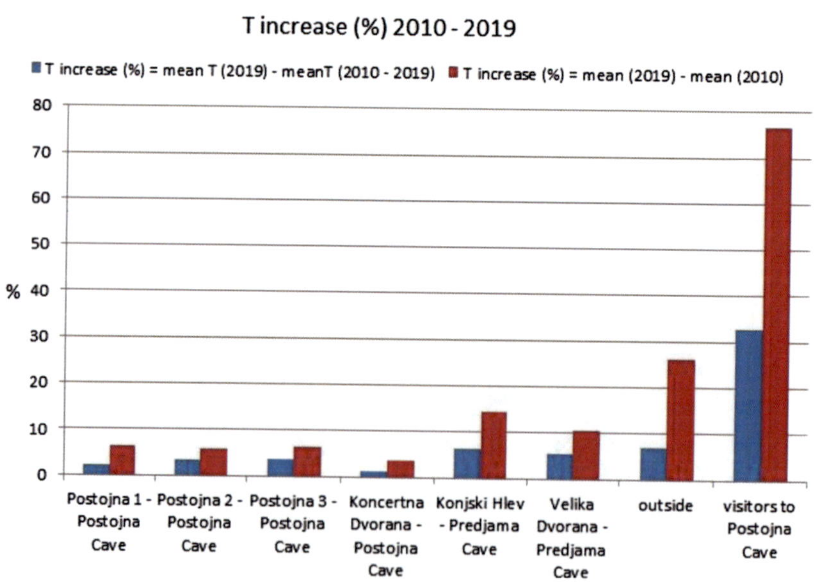

Figure 9 Air temperature and visitor numbers increase (%) for the period 2010–19 (Koncertna Dvorana for 2011–19).

Postojna 2, +6.4% for Postojna 3 and +3.6% for Koncertna Dvorana). There was a 76% increase in the number of visitors to Postojna Cave between 2010 and 2019. Also, the outside air temperature shows a high increase (+26%) in 2019 compared with 2010. The increase in the air temperature at the Predjama Cave sites reflects the increase in the outside air temperature and is not connected with the presence of visitors. All Postojna Cave temperature monitoring sites are situated 2–3 km deep inside the cave, while the Predjama Cave sites are closer to the cave entrance (50–200 m), which may mean that the inner Postojna Cave passages, situated in deep karst massive, have been more isolated and resistant to temperature entry for the calculated period of 2010–19, than the Predjama Cave sites.

If we compare the mean annual cave air temperature with the mean annual outside air temperature we can see that the most visited parts of Postojna Cave are overheated compared with the outside conditions and Predjama Cave is hypothermic compared with the outside conditions (Fig. 10). Even if Predjama Cave has been heated more than Postojna Cave since 2010 (Fig. 9) due to the increase in outside air temperature it remains colder than Postojna Cave, which illustrates the accumulated heat deep

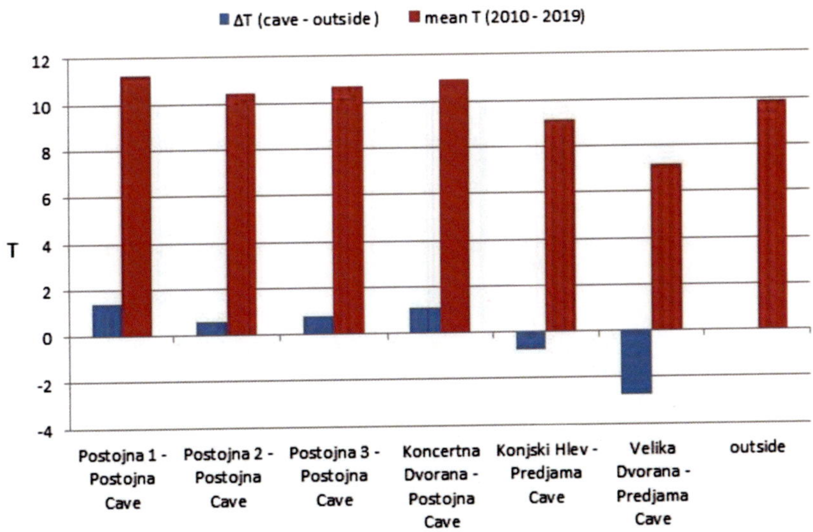

Figure 10 Mean air temperature in °C for the period 2010–19, and the difference between mean cave air temperature and outside air temperature for the period 2010–19 (Koncertna Dvorana for 2011–19).

inside the Postojna Cave system (probably existing before 2010), partly due to the increase in outside air temperature and also due to the impact of visitors, including heat from people and lights.

Being a show cave for more than 200 years, the anthropogenic impacts on the Postojna Cave microclimate due to visitors have been detected in changes in air temperature and carbon dioxide levels. These are mostly short-term impacts because after visitors leave the cave, both parameters return to their previsit values. On the other hand, the most overheated cave site is almost free of tourists—Otoška Jama (Fig. 11A and B) with a mean air

Figure 11 Otoška Jama: (A) entrance, (B) cave passage. *(Photo: S. Šebela.)*

Figure 12 Since 1872 a cave train has been in operation at Postojna Cave. *(Photo: S. Šebela.)*

Figure 13 Postojna Cave main entrance. *(Photo: S. Šebela.)*

temperature 12.51°C (2018–19). It was found that the Otoška Jama 2 site is the warmest monitoring site in the whole Postojna Cave system, as it has a 1.95°C higher mean annual air temperature than the outside mean annual air temperature. This is ascribed to the natural microclimate conditions, especially with the fact that a lot of heat is transmitted to the air from the underground River Pivka which causes the characteristics of a warm-air trap cave.

On May 29, 2019, the Postojna Cave (Figs. 12 and 13) celebrated 39 million visitors since 1819 when official visits to the newly discovered cave parts were started. There have been four significant drops in the number of visitors since 1819. The first drop was related to the First World War, the second drop to the Second World War, and the third drop to the independence of Slovenia in 1991, and the politically unstable conditions in the surrounding region. The last and fourth drop has been due to the COVID-19 Coronavirus pandemic, which started in March 2020. In 2020 there were about 85% less visitors than in 2019.

References

Agapito, G., 1823. Le Grotte ed altri notevoli oggetti nelle vicinanze di Trieste. In: Dalla Tipografia di Antonio Strauss, a spese di Paolo Schubart in Trieste, Vienna (in Italian).

Anelli, F., 1941–44. Osservazioni di meteorologia ipogea nelle grotte di Castel Lueghi presso Postumia (N. 117 V.G.). Le Grotte d'Italia 2a (V), pp. 5–34 (in Italian).

ARSO, 2020. Lidar Data. Available from: http://gis.arso.gov.si/evode/profile.aspx?id=atlas_voda_Lidar@Arso.

Badino, G., 1995. Fisica del Clima Sotterraneo, Memorie IIS 7 (II), p. 137 (in Italian).

Badino, G., 1999. Il clima sotterraneo. In: La Venta Association – Members Writtings E-Archives, pp. 1–49. Available from: www.laventa.it.

Badino, G., 2004. Cave temperatures and global climatic change. Int. J. Speleol. 33 (1/4), 103–114.

Badino, G., 2010. Underground meteorology-"what's the weather underground?". Acta Carsol. 39 (3), 427–448. https://doi.org/10.3986/ac.v39i3.74.

Badino, G., Chignola, R., 2019. Fluctuations of atmospheric pressure and sound of underground Karst systems: the Antro del Corchia Case (Apuane Alps, Italy). Front. Earth Sci. 7 (147), 1–6. https://doi.org/10.3389/feart.2019.00147.

Baker, A.C., 2014. The Jenolan environmental monitoring programe. Proc. Linn. Soc. N. S. W. 136, 19–34. Available from: http://escholarship.library.usyd.edu.au/jounals/index.php/LIN.

Baker, A., Genty, D., 1998. Environmental pressures on conserving cave spleleothems: effects of changing surface land use and increased tourism. J. Environ. Manag. 53, 165–175.

Baldini, J.U.L., McDermott, F., Fairchild, I.J., 2002. Structure of the 8200-year cold event revealed by a speleothem trace element record. Science 296, 2203–2206.

Bavdek, A., 2003. Arheološki nadzor v Postojnski jami. Naše jame 45, pp. 143–145 (in Slovene).

Bavdek, A., Čuk, A., 2015. The Predjama Treasure Cache. Institute Znanje Postojna, OU Notranjska Museum Postojna, Postojna.

Bavdek, A., Josipović, D. (Eds.), 2017. Betalov spodmol. Notranjski muzej Postojna, Postojna.

Becker, A., Ferry, M., Monecke, K., Schnellmann, M., Giardini, D., 2005. Multiarchive paleoseismic record of late Pleistocene and Holocene strong earthquakes in Switzerland. Tectonophysics 400, 153–177.

Blatnik, M., Culver, D.C., Gabrovšek, F., Knez, M., Kogovšek, B., Kogovšek, J., Liu, H., Mayaud, C., Mihevc, A., Mulec, J., Aljančič, M., Otoničar, B., Petrič, M., Pipan, T., Prelovšek, M., Ravbar, N., Shaw, T.R., Slabe, T., Šebela, S., Zupan Hajna, N., 2020. Karstology in the Classical Karst (Advances in Karst Science). Springer, Cham. https://doi.org/10.1007/978-3-030-26827-5.

Boes, C., Ek, C., Faber, S., Kies, A., Massen, F., Schintgen, G., Sinner, E., Waringo, G., 1997. The Moestroff Cave, A Study on the Geology and Climate of Luxembourg's Largest Maze Cave. Centre de Recherches Public Centre Universitaire, Luxembourg.

Bourges, F., Genthon, P., Mangin, A., D'Hulst, D., 2006. Microclimates of l'aven d'Orgnac and other French limestone caves (Chauvet, Esparros, Marsoulas). Int. J. Climatol. 26 (1), 1651–1670.

Bourges, F., Genthon, P., Genty, D., Lorblanchet, M., Mauduit, E., D'Hulst, D., 2014. Conservation of prehistoric caves and stability of their inner climate: lessons from

Chauvet and other French caves. Sci. Total Environ. 493, 79–91. https://doi.org/10.1016/j.scitotenv.2014.05.137.
Brenčič, M., 2018. Analysis of annual visits to Postojnska jama – historical and statistical sciences approach. In: Prelovšek, M. (Ed.), Show Caves and Science: Abstracts & Guide Book. 18–22 June 2018. Postojna, Založba ZRC, Ljubljana, p. 104.
Brodar, S., 1951. Otoška jama, paleolitska postaja. Razprave SAZU 1, pp. 203–242.
Brodar, S., 1966. Pleistozäne Sedimente und Paläolithische Fundstellen in der Postojnska jama. Acta Carsol. 4, 1–84.
Čar, J., Šebela, S., 2001. Karst characteristics of thrust contact limestone-dolomite near Predjama. Acta Carsol. 30/2, 141–156.
Carnera, L., 1933. La stazione dei pendoli orizzontali nelle R.R. Grotte di Postumia. Bollettino di Geodesia e Geofisica 9–10 (XII), 1–13.
Carrasco, F., Vadillo, I., Liñán, C., Andreo, B., Durán, J.J., 2002. Control of environmental parameters for management and conservation of Nerja Cave (Malaga, Spain). Acta Carsol. 31 (1), 105–122. https://doi.org/10.3986/ac.v31i1.407.
Cave Register of the Republic of Slovenia. Available from: https://www.katasterjam.si/.
Cigna, A.A., Buri, E., 2000. Development, management and economy of show caves. Int. J. Speleol. 29 (B), 1–27.
Cigna, A.A., Forti, P., 2013. Caves: the most important geotouristic feature in the World. Tourism and Karst Areas 6 (1), 9–26.
Covington, M.D., Perne, M., 2015. Consider a cylindrical cave: a physicist's view of cave and karst science. Acta Carsol. 44 (3), 363–380. Available from: https://ojs.zrc-sazu.si/carsologica/article/view/1925.
Covington, M.D., Luhmann, A.J., Gabrovšek, F., Saar, M.O., Wicks, C.M., 2011. Mechanisms of heat exchange between water and rock in karst conduits. Water Resour. Res. 47 (W10514), 1–18. https://doi.org/10.1029/2011WR010683.
Crestani, G., Anelli, F., 1939. Ricerche di meteorologia ipogea delle grotte di Postumia. Istituto poligrafico dello stato Libreria, Roma (in Italian).
Cropley, J.B., 1965. Influence of surface conditions on temperatures in large cave systems. Natl. Speleol. Soc. Bull. 27 (1), 1–10.
Čuk, A., 2008. Postojna Cave. Turizem KRAS, destinacijski Management d.d., ČukGraf, Postojna.
Culver, D.C., Pipan, T., 2009. The Biology of Caves and Other Subterranean Habitats (Biology of Habitats). Oxford University Press, Oxford, New York.
Culver, D.C., Pipan, T., 2014. Shallow Subterranean Habitats: Ecology, Evolution, and Conservation. Oxford University Press, New York.
Culver, D.C., Debevec, B., Knez, M., Kovačič, G., Kranjc, A., Mulec, J., Pipan, T., Prelovšek, M., Ravbar, N., Semeja, A., Slabe, T., Šebela, S., Zupan Hajna, N., 2012. Karstology and Development Challenges on Karst 2, Construction, Tourism, Ecology, Protection. ZRC Publishing, Carsologica 14, Postojna-Ljubljana.
de Freitas, C.R., 2010. The role and importance of cave microclimate in the sustainable use and management of show caves. Acta Carsol. 39 (3), 477–489. https://doi.org/10.3986/ac.v39i3.77.
Debevec, V., Peric, B., Šturm, S., Zorman, T., Jovanovič, P., 2018. Škocjan caves, Slovenia: an integrative approach to the management of a world heritage site. In: Parise, M., Gabrovšek, F., Kaufman, G., Ravbar, N. (Eds.), Advances in Karst Research: Theory, Fieldwork and Applications. Geological Society, Special Publications, 466, London, pp. 411–429. https://doi.org/10.1144/SP466.14.
Dirjec, J., Turk, I., 1987. Postojna – Postojnska jama. Varstvo spomenikov 29, pp. 232–233.
Domicelj, A., January 15, 1990. Huda nesreča na Notranjskem. Dom in Svet. Year XIII, no 2, Ljubljana ((in Slovene).

Domínguez-Villar, D., Lojen, S., Krklec, K., Baker, A., Fairchild, I.J., 2015. Is global warming affecting cave temperatures? Experimental and model data from a paradigmatic case study. Clim. Dynam. 45, 569–681. https://doi.org/10.1007/s00382-014-2226-1.

Edenborn, H.M., Sams, J.I., Kite, J.S., 2012. Thermal regime of a cold air trap in central Pennsylvania, USA: the trough creek ice mine. Permafr. Periglac. Process. 23, 187–195. https://doi.org/10.1002/ppp.1742.

Edinost, 1926. Postojna, sedem dni potresa, 8.1.1926, 8 (in Slovene).

Elliotte, W.R., 2012. Protecting caves and cave life. In: White, W.B., Culver, D.C. (Eds.), Encyclopedia of Caves. Academic Press, pp. 624–634.

Faimon, J., Štelcl, J., Kubešová, S., Zimák, J., 2003. Environmentally acceptable effect of hydrogen peroxide on cave "lamp-flora", calcite speleothems and limestones. Environ. Pollut. 122, 417–422.

Faimon, J., Stelcl, J., Sas, D., 2006. Anthropogenic CO_2-flux into cave atmosphere and its environmental impact: a case study in the Císařská Cave (Moravian Karst, Czech Republic). Sci. Total Environ. 369 (1–3), 231–245. https://doi.org/10.1016/j.scitotenv.2006.04.006.

Faimon, J., Troppová, D., Baldík, V., Novotný, R., 2012. Air circulation and its impact on microclimatic variables in the Císařská Cave (Moravian Karst, Czech Republic). Int. J. Climatol. 32, 599–623. https://doi.org/10.1002/joc.2298.

Faimon, J., Lang, M., Geršl, M., Sracek, O., Bábek, O., 2020. The »breathing spots« in karst areas—the sites of advective exchange of gases between soils and adjacent underground cavities. Theor. Appl. Climatol. 142, 85–101. https://doi.org/10.1007/s00704-020-03280-7.

Fatur, S., 1975. Partizanska akcija v Postojnski jami. In: Fatur, S. (Ed.), Ljudje in kraji ob Pivki, Kulturna skupnost Postojna, pp. 298–301. Postojna (in Slovene).

Fidel, E., 2020. Jamarji v jamskem sistemu pod Predjamskim gradom odkrili nove rove. Postojnski prepih Newspaper (107), 13. Postojna (in Slovene).

Ford, D.C., Williams, P., 1989. Karst Geomorphology and Hydrology. Chapman and Hall, London.

Ford, D.C., Williams, P., 2007. Karst Hydrogeology and Geomorphology. Wiley and sons, Chichester.

Gabrovšek, F., Mihevc, A. (Eds.), 2009. 17th International Karstological School »Classical Karst«, Guide Book & Abstracts, 15–20 June 2009. Postojna, Založba ZRC, Postojna, p. 92.

Gabrovšek, F., Grašič, B., Božnar, M., Mlakar, P., Udén, M., Davies, E., 2014. Karst show caves – how DTN technology as used in space assists automatic environmental monitoring and tourist protection – experiment in Postojna Cave. Nat. Hazards Earth Syst. Sci. 14 (2), 443–457. https://doi.org/10.5194/nhess-14-443-2014.

Gams, I., 1968. Über die Faktoren, die Intensität der Sintersedimentation bestimmen. In: Bohinec, V. (Ed.), Proceedings of the 4th International Congress of Speleology in Yugoslavia, 12–26 September 1965, Dubrovnik. Speleological Society of Yugoslavia, Ljubljana, pp. 107–115.

Gams, I., 1970. Luftzirkulation als bestandteil de höhlenmilieus am beispiel der Höhle von Postojna. In: Karamani, Z., Manaković, D. (Eds.), Congrès Yougoslave de spéléologie cinquième session, 15–20 September 1968, Skopje, Ohrid, Speleološki sojuz na Jugoslavija. Prirodno-matematički fakultet, Skopje, pp. 99–111.

Gams, I., 1974. Concentration of CO_2 in the caves in relation to the air circulation (in the case of Postojna Cave). Acta Carsol. 6, 183–192.

Gams, I., 1986. Some methods of the research of cave processes. Naše Jame 28, 32–38 (in Slovene with English Summary).

Gams, I., 2002. Vpliv vdiranja zimskega mraza na jamsko okolje. Postojnska jama in primorske turistične jame. Proteus 65 (2), 102–111 (in Slovene).

Gams, I., 2003. Kras v Sloveniji v prostoru in času. Založba ZRC, Ljubljana (in Slovene with English summary).

Gams, I., 2004. Kras v Sloveniji v prostoru in času. Založba ZRC, Ljubljana (in Slovene with English summary).

Gams, I., Kogovšek, J., 1998. The dynamics of flowstone deposition in the caves Postojnska, Planinska, Taborska and Škocjanske. Acta Carsol. 27 (1), 299–324.

Garofano, M., Govoni, D., 2012. Underground geotourism: a historic and economic overview of show caves and show Mines in Italy. Geoheritage 4, 79–92. https://doi.org/10.1007/s12371-012-0055-3.

Gázquez, F., Quindós-Poncela, Sainz-Fernández, C., Fernández-Villar, A., Fuente-Merino, I., Celaya-Gonzalez, S., 2016. Spatiotemporal distribution of δ13C-CO_2 in a shallow cave and its potential use as indicator of anthropic pressure. J. Environ. Manag. 180, 421–432. https://doi.org/10.1016/j.jenvman.2016.05.078.

Gilli, E., 1999. Evidence of palaeoseismicity in a flowstone of the Observatoire cave (Monaco). Geodin. Acta 12 (3–4), 159–168.

Glažar, S., 2006. Strela in potres v Postojnski jami. Naše Jame 46, 118–119 (in Slovene).

Gregorič, A., Zidanšek, A., Vaupotič, J., 2011. Dependence of radon levels in Postojna Cave on outside air temperature. Nat. Hazards Earth Syst. Sci. 11 (5), 1523–1528. https://doi.org/10.5194/nhess-11-1523-2011.

Gregorič, A., Vaupotič, J., Gabrovšek, F., 2013. Reasons for large fluctuation of radon and CO_2 levels in a dead-end passage of a karst cave (Postojna Cave, Slovenia). Nat. Hazards Earth Syst. Sci. 13 (2), 287–297. https://doi.org/10.5194/nhess-13-287-2013.

Gregorič, A., Vaupotič, J., Šebela, S., 2014. The role of cave ventilation in governing cave air temperature and radon levels (Postojna Cave, Slovenia). Int. J. Climatol. 34, 1488–1500. https://doi.org/10.1002/joc.3778.

Grgić, I., Iskra, I., Podkrajšek, B., Debevec Gerjevič, V., 2014. Measurements of aerosol particles in the Škocjan Caves, Slovenia. Environ. Sci. Pollut. Res. Int. 21 (3), 1915–1923. https://doi.org/10.1007/s11356-013-2080-4.

Gruber, T., 1781. Briefe hydrographischen und physikalischen Inhalts aus Krain, pp. 1–159. Laibach (in German).

Guerrier, B., Doumenc, F., Roux, A., Mergui, S., Jeannin, P.-Y., 2019. Climatology in shallow caves with negligible ventilation: heat and mass transfer. Int. J. Therm. Sci. 146, 106066. https://doi.org/10.1016/j.ijthermalsci.2019.106066.

Habe, F., 1966. Katastrofalne poplave pred našimi turističnimi jamami. Naše Jame 8, 45–54 (in Slovene).

Habe, F., 1970. Die höhlenwelt von Predjama. Acta Carsol. 5 (1), 5–94.

Habe, F., 1986. The Postojna caves and Other Tourist Caves in Slovenia. Tone Tomšič Printing House, Ljubljana.

Habe, F., 1987. Predjama. The Castle and the Cave. "Postojnska Jama". Tourist and Hotel Organization, Postojna.

Habe, F., 1996. Mlini in žage na vodni pogon na Pivki in Planinskem polju nekoč in danes. Tiskarna Lotos, Postojna (in Slovene).

Hacquet, B., 1778–1789. Oryctographia Carniolica, vol. 1, pp. 128–129 (Leipzig).

Hoyos, M., Soler, V., Cañaveras, J.C., Sánchez-Moral, S., Sanz-Rubio, E., 1998. Microclimatic characterization of a karstic cave: human impact on microenvironmental parameters of a prehistoric rock art cave (Candamo Cave, northern Spain). Environ. Geol. 33 (4), 231–242.

Jurkovšek, B., Biolchi, S., Furlani, S., Kolar-Jurkovšek, T., Zini, L., Jež, J., Tunis, G., Bavec, M., Cucchi, F., 2016. Geology of the classical karst region (SW Slovenia–NE Italy). J. Maps 12, 352–362. https://doi.org/10.1080/17445647.2016.1215941.

Kagan, E.J., Agnon, A., Bar-Matthews, M., Ayalon, A., 2005. Dating large infrequent earthquakes by damaged cave deposits. Geology 33 (4), 261–264.

Kalnay, E., et al., 1996. The NCEP/NCAR 40-year reanalysis Project. Bull. Am. Meteorol. Soc. 77 (3), 437–472. https://doi.org/10.1175/1520-0477(1996)077<0437:TNYRP>2.0.CO;2.

Kaufmann, G., Gabrovšek, F., Turk, J., 2016. Modelling flow of subterranean Pivka river in Postojnska jama, Slovenia. Acta Carsol. 45 (1), 57–70.

Killing-Heinze, M., Pflisch, A., Furian, W., Allison, S., 2017. The importance of air temperature as a key parameter to identify climatic processes inside Carlsbad Cavern, New Mexico, USA. J. Cave Karst Stud. 79 (3), 153–167. https://doi.org/10.4311/2014IC0119.

Kogovšek, J., 2010. Characteristics of Percolation Through the Karst vadose Zone. Carsologica 10. Založba ZRC, Postojna–Ljubljana.

Komac, B., Natek, K., Zorn, M., 2008. Geographical aspects of floods in Slovenia. In: Geografija Slovenije 20. Založba ZRC, Ljubljana (in Slovene with English abstract).

Korošec, P., 1975. Eneolitik Slovenije. Arheološki vestnik 24 (1973), pp. 84–104 (in Slovene).

Krajnc, B., Ferlan, M., Ogrinc, N., 2017. Soil CO_2 sources above a subterranean cave: Pisani rov (Postojna Cave, Slovenia). J. Soils Sediments 17 (7), 1883–1892. https://doi.org/10.1007/s11368-016-1543-x.

Kranjc, A., 1983. Dinamika odpadanja sige v Golobji luknji, Predjama. Acta Carsol. 11, 99–116.

Kranjc, A., 1998. Hydrological characteristics of Škocjanske jame. Naše Jame 40, 28–33 (in Slovene with English summary).

Kranjc, A., 2004. Short history of cave tourism in Slovenia. In: Zupan Hajna, N. (Ed.), IV° I.S.C.A. International Congress, 21–27 October 2002. Postojna, Postojna Caves, Ljubljana, pp. 55–60.

Kranjc, A., 2010. Vilenica. DEDI – digitalna enciklopedija naravne in kulturne dediščine na Slovenskem. Available from: http://www.dedi.si/dediscina/85-vilenica.

Kranjc, A., Malečkar, F., 1988. Postojnska Jama, 170 Let odkrivanja, raziskovanja in turističnega razvoja. Vodnik 10. Tiskarna Tone Tomšič, Postojna.

Kranjc, A., Opara, B., 2002. Temperature monitoring in Škocjanske jame. Acta Carsol. 31 (1), 85–96.

Kranjc, A., Kariž, S., Paternost, S., Polak, S., 2007. Postojnska jama vodnik. In: ČukGraf Postojna, Postojnska jama turizem d.d. Postojna.

Lang, M., Faimon, J., Godissart, J., Ek, C., 2017a. Carbon dioxide seasonality in dynamically ventilated caves: the role of advective fluxes. Theor. Appl. Climatol. 129, 1355–1372. https://doi.org/10.1007/s00704-016-1858-y.

Lang, M., Faimon, J., Kejíková, S., 2017b. The impact of door opening on CO_2 levels: a case study from the Balcarka Cave (Moravian Karst, Czech Republic). Int. J. Speleol. 46 (3), 345–358. https://doi.org/10.5038/1827-806X.46.3.2100.

Liñán, C., del Rosal, Y., 2015. Natural ventilation of karstic caves: new data on the Nerja Cave (Malaga, S of Spain). In: Andreo, B., Carrasco, F., Durán, J., Jiménez, P., LaMoreaux, J. (Eds.), Hydrogeological and Environmental Investigations in Karst Systems. Environmental Earth Sciences, vol. 1. Springer, Berlin, Heidelberg. https://doi.org/10.1007/978-3-642-17435-3_57.

Liñán, C., Vadillo, I., Carrasco, F., 2008. Carbon dioxide concentration in air within the Nerja Cave (Malaga, Andalusia, Spain). Int. J. Speleol. 37 (2), 99–106. https://doi.org/10.5038/1827-806X.37.2.2.

Liñán, C., del Rosal, Y., Carrasco, F., Vadillo, I., Benavente, J., Ojeda, L., 2018. Highlighting the importance of transitional ventilation regimes in the management of Mediterranean show caves (Nerja-Pintada system, southern Spain). Sci.Total Environ. 631–632, 1268–1278. https://doi.org/10.1016/j.scitotenv.2018.02.3040.

Liñán, C., Ojeda, L., Benavente, J., del Rosal, Y., Vadillo, I., Carrasco, F., 2020. Coupling air temperature records and gravimetric data to interpret ventilation patterns in a Mediterranean karstic system (Nerja-Pintada caves, southern Spain). Sci. Total Environ. 730, 139147. https://doi.org/10.1016/j.scitotenv.2020.139147.

Lütscher, M., Jeannin, P.-Y., Haeberli, W., 2005. Ice caves as an indicator of winter climate evolution: a case study from the Jura Mountains. Holocene 15 (7), 982–993. https://doi.org/10.1191/0959683605hl872ra.

Luhmann, A.J., Covington, M.D., Myre, J.M., Perne, M., Jones, S.W., Alexander Jr., E.C., Saar, M.O., 2015. Thermal damping and retardation in karst conduits. Hydrol. Earth Syst. Sci. 19, 137–157. https://doi.org/10.5194/hess-19-137-2015.

Lütscher, M., Jeannin, P.Y., 2004. Temperature distribution in karst systems: the role of air and water fluxes. Terra. Nova 16, 344–350. https://doi.org/10.1007/s00704-016-1858-y.

Lăzărescu, H., Simionica, I., Hoteteu, M., Mirescu, L., 2014. Speleotherapy – modern bio-medical perspectives. J. Med. Life 7 (2), 76–79.

Mandić, M., Mihevc, A., Leis, A., Krajcar Bronić, I., 2013. Concentration of stable carbon isotopic composition of CO_2 in cave air of Postojnska Jama, Slovenia. Int. J. Speleol. 42 (3), 279–287. https://doi.org/10.5038/1827-806X.42.3.11.

Mulec, J., 2014. Human impact on underground cultural and natural heritage sites, biological parameters of monitoring and remediation actions for insensitive surfaces: case of Slovenian show caves. J. Nat. Conserv. 22 (2), 132–141. https://doi.org/10.1016/j.jnc.2013.10.001.

Mulec, J., Kosi, G., 2009. Lampenflora algae and methods of growth control. J. Cave Karst Stud. 71 (2), 109–115.

Mulec, J., Oarga-Mulec, A., Šturm, S., Tomazin, R., Matos, T., 2017. Spacio-temporal distribution and tourist impact on airborne bacteria in a cave (Škocjan Caves, Slovenia). Diversity 9 (3), 1–14. https://doi.org/10.3390/d9030028.

Muri, G., Jovičić, A., Mihevc, A., 2013. Source assessment of deposited particles in a Slovenian show cave (Postojnska jama): evidence of long-lasting anthropogenic impact. Int. J. Speleol. 42, 225–233.

Nagel, J.A., 1748. Beschreibung deren auf allerhöchsten Befehl Ihro Röm. Kayserl. Köngl. Maytt: Francisci I. in dem Herzogthume Crain befindlich Seltenheiten der Natur. Nationalbibliothek, manuscript Nr. 7854, Wien (in German).

Obu, J., Košutnik, J., Overduin, P.P., Boike, J.A., Blatnik, M., Zwieback, S., Gostinčar, P., Mihevc, A., 2018. Sorted patterned ground in a karst cave, Ledenica pod Hrušico, Slovenia. Permafr. Periglac. Process. 29 (2), 121–130. https://doi.org/10.1002/ppp.1970.

Ogrin, D., 2015. Long-term air temperature changes in Ljubljana (Slovenia) in comparison to Trieste (Italy) and Zagreb (Croatia). Morav. Geogr. Rep. 23 (3), 17–26. https://doi.org/10.1515/mgr-2015-0014.

Palmer, A.N., 2007. Cave Geology. Cave Books, Dayton, OH, USA.

Parmesan, C., Yohe, G., 2003. A globally coherent fingerprint of climate change impacts across natural systems. Nature 421, 37–42. Available from: www.nature.com/nature.

Peric, B., Hribar, M., 2009. Škocjanske jame. DEDI – Digital Encyclopedia of Natural and Cultural Heritage of Slovenia. http://www.dedi.si/dediscina/8-skocjanske-jame.

Peyraube, N., Lastennet, R., Denis, A., Malaurent, P., Houillon, N., Villanueva, J.D., 2018. Determination and quantification of major climatic parameters influencing the CO_2 of Lascaux Cave. Theor. Appl. Climatol. 133, 1291–1301. https://doi.org/10.1007/s00704-017-2255-x.

Pflitsch, A., Wiles, M., Horrocks, R., Piasecki, J., Ringeis, J., 2010. Dynamic climatologic processes of barometric cave systems using the example of Jewel Cave and Wind Cave

in South Dakota, USA. Acta Carsol. 39 (3), 449–462. https://doi.org/10.3986/ac.v39i3.75.
Pipan, T., Petrič, M., Šebela, S., Culver, D.C., 2019. Analyzing climate change and surface-subsurface interactions using the Postojna Planina Cave System (Slovenia) as a model system. Reg. Environ. Change 19 (2), 379–389. https://doi.org/10.1007/s10113-018-1349-z.
Politi, I., 1939–40. Sui fenomeni di annerimento delle formazioni calcitiche e calcaree nelle Grotte di Postumia. Le Grotte d'Italia, serie 2a, vol. IV, pp. 18–23 (in Italian).
Poljak, M., Živčić, M., Zupančič, P., 2000. The seismotectonic characteristics of Slovenia. Pure Appl. Geophys. 157, 37–55.
Postpischl, D., Agostini, S., Forti, P., Quinif, Y., 1991. Palaeoseismicity from karst sediments: the »Grotta del Cervo« cave case study (central Italy). Tectonophysics 193, 33–44.
Prelovšek, M., Šebela, S., Turk, J., 2018. Carbon dioxide in Postojna Cave (Slovenia): spatial distribution, seasonal dynamics and evaluation of plausible sources and sinks. Environ. Earth Sci. 77 (7), 1–15. https://doi.org/10.1007/s12665-018-7459-6.
Rakovec, J., 2020. Impact of visits on microclimate of caves – an analytical approach. Acta Carsol. 49 (1), 97–107. https://doi.org/10.3986/ac.v49i1.8721.
Ravbar, N., Košutnik, J., 2014. Variations of karst underground air temperature induced by various factors (Cave of Županova jama, Central Slovenia). Theor. Appl. Climatol. 116, 327–341. https://doi.org/10.1007/s00704-013-0955-4.
R. R. Grotte demaniali di Postumia, 1928. Relazione del consiglio d'amministrazione alle loro eccellenze i ministri dell'economia nazionale e delle finanze sull'andamento dell'azienda dal 1° luglio 1926 al 31 dicembre 1927. VI, 12, Postumia (in Italian).
R. R. Grotte demaniali di Postumia, 1938. Relazione del consiglio d'amministrazione alle loro eccellenze i ministri dell'economia nazionale e delle finanze sull'andamento dell'azienda dal 1° gennaio al 31 dicembre 1937. XVI, Postumia (in Italian).
Ribarič, V., 1982. Seismicity of Slovenia – Catalogue of Earthquakes (792 A.D. – 1981). SZ SRS Publication, Ljubljana, pp. 1–650. Ser. A, No. 1–1.
Schmidl, A., 1854. Die Grotten und Höhlen von Adelsberg, Lueg, Planina und Laas. Wien: Akademie der Wissenschaften. Wien, V–VIII, pp. 3–314 (in German).
Schönleben, J.L., 1681. Carniola antiqua et nova. Ed. I, Cap. Iv, Laibach.
Šebela, S., 1998. Tectonic Structure of Postojnska Jama Cave System. Zbirka ZRC 18, Založba ZRC, Ljubljana. https://doi.org/10.3986/961618265X.
Šebela, S., 2008. Broken speleothems as indicators of tectonic movements. Acta Carsol. 37 (1), 51–62.
Šebela, S., 2010. Accesses from the Surface to the Postojna Cave System. Annales, Ser. Hist. Nat. 20, 55–64.
Šebela, S., 2011. Exceptional natural events in Postojna and Predjama Cave systems. Annales-Anali Istrske Mediteranske Studije-Ser. Hist. Nat. 21 (1), 87–94 (in Slovene with English abstract).
Šebela, S., 2012. Historic inscriptions in Predjama cave system and high floods in 2010. Acta Carsol. 41 (2–3), 297–303.
Šebela, S., 2019. Postojna – Planina Cave system, Slovenia. In: White, W., Culver, D., Pipan, T. (Eds.), Encyclopedia of Caves, third ed. Academic Press, pp. 812–821. https://doi.org/10.1016/B978-0-12-814124-3.00098-4.
Šebela, S., Čar, J., 1991. Geological setting of collapse chambers in Vzhodni Rov in Predjama Cave. Acta Carsol. 20, 205–222.
Šebela, S., Turk, J., 2011a. Local characteristics of Postojna Cave climate, air temperature, and pressure monitoring. Theor. Appl. Climatol. 105, 371–386. https://doi.org/10.1007/s00704-011-0397-9.

Šebela, S., Turk, J., 2011b. Air temperature characteristics of Postojna and Predjama cave systems. Acta Geogr. Slov. 51 (1), 43–64. https://doi.org/10.3986/AGS51102.
Šebela, S., Turk, J., 2014a. Natural and anthropogenic influences on the year-round temperature dynamics of air and water in Postojna show cave, Slovenia. Tourism Manag. 40, 233–243. https://doi.org/10.1016/j.tourman.2013.06.011.
Šebela, S., Turk, J., 2014b. Sustainable use of the Predjama Cave (Slovenia) and possible scenarios related to anticipated major increases in tourist numbers. Tourism Manag. Perspect. 10, 37–45. https://doi.org/10.1016/j.tmp.2014.01.002.
Šebela, S., Turk, J., 2018. Črna Jama as a cold air trap cave within Postojna Cave, Slovenia. Theor. Appl. Climatol. 134 (3–4), 741–751. https://doi.org/10.1007/s00704-017-2304-5.
Šebela, S., Vaupotič, J., Košťák, B., Stemberk, J., 2010. Direct measurement of present-day tectonic movement and associated radon flux in Postojna cave, Slovenia. J. Caves Karst Stud. 72 (1), 21–34. https://doi.org/10.4311/jcks2009es0077.
Šebela, S., Prelovšek, M., Turk, J., 2013. Impact of peak period visits on the Postojna cave (Slovenia) microclimate. Theor. Appl. Climatol. 111, 51–64. https://doi.org/10.1007/s00704-012-0644-8.
Šebela, S., Turk, J., Pipan, T., 2015a. Cave micro-climate and tourism: towards 200 years (1819–2015) at Postojnska jama (Slovenia). Cave Karst Sci. 42 (2), 78–85. Available from: http://bcra.org.uk/pub/candks/index.html?j.
Šebela, S., Miler, M., Skobe, S., Torkar, S., Zupančič, N., 2015b. Characterization of black deposits in karst caves, examples from Slovenia. Facies 61 (2), 1–13. https://doi.org/10.1007/s10347-015-0430-z.
Šebela, S., Zupančič, N., Miler, M., Grčman, H., Jarc, S., 2017. Evidence of Holocene surface and near-surface palaeofires in karst caves and soils. Palaeogeogr. Palaeoclimatol. Palaeoecol. 485, 224–235. https://doi.org/10.1016/j.palaeo.2017.06015.
Shaw, T.R., 2003. Early electric lighting in caves – Postojnska Jama, Slovenia, 1883 – 1929. Acta Carsol. 32 (1), 189–204.
Shaw, T.R., 2008. Foreign Travellers in the Slovene Karst: 1486 – 1900. Založba ZRC, Ljubljana.
Shaw, T.R., Čuk, A., 2002. Royal and other noble visitors of Postojnska jama 1819 – 1945. Acta Carsol. 31 (1), 5–106.
Shaw, T.R., Čuk, A., 2015. Slovene Karst and Caves in the Past. ZRC Publishing, Ljubljana.
Smithson, P.A., 1991. Inter-relationship between cave and outside air temperatures. Theor. Appl. Climatol. 44, 65–73.
Steinberg, A., 1758. Gründliche Nachricht von dem in dem Inner-Krain gelegenen Zirknitzer See. Laybach, pp. 200–201 (in German).
Turk, J., 2011. Geoarchaeological research into Paleolithic cave sites as a source of palaeoenvironmental data. Arheol. Vestn. 62, 9–31.
Valvasor, J.W., 1689. Die Ehre dess Herzogthums Crain. I. Th. Wolfgang Moritz Endter Buchhändlern, Laibach-Nürnberg, p. 696 (in German).
Vaupotič, J., 2008. Nanosize radon short-lived decay products in the air of the Postojna Cave. Sci. Total Environ. 393 (1), 27–38.
Vaupotič, J., Dujmović, P., Kobal, I., 1998. Radiation doses due to radon and progency in Postojna cave. Acta Carsol. 27 (1), 395–406.
Vaupotič, J., Csige, I., Radolić, V., Hunyadi, I., Planinić, J., Kobal, I., 2001. Methodology of radon monitoring and dose estimates in Postojna cave, Slovenia. Health Phys. 80, 142–147.
Vercelli, F., 1931. Il regime termico nelle Grotte di San Canziano. Grotte Italia 5 (2), 49–62 (in Italian).

Villar, E., Fernandez, P.L., Quindos, L.S., Solana, J.R., Soto, J., 1984. Air temperatures and air interchanges at Altamira Cave (Spain). Cave science. Trans. Br. Cave Res. Assoc. 11 (2), 92–98.

Vokal, B., Obelić, B., Genty, D., Kobal, I., 1999. Chemistry measurements of dripping water in Postojna Cave. Acta Carsol. 28 (1), 305–321.

Walters, R., Zupan Hajna, N., 2020. 3D laser scanning of the natural caves: example of Škocjanske Jame. Geod. Vestn. 64 (1), 89–103. https://doi.org/10.15292/geodetski-vestnik.2020.01.89-103.

White, W.B., 2007. Cave sediments and paleoclimate. J. Cave Karst Stud. 69 (1), 76–93.

Williams, K.E., McKay, C.P., 2015. Comparing flow-through and static ice cave models for Shoshone Ice Cave. Int. J. Speleol. 44, 115–123. https://doi.org/10.5038/1827-806X.44.2.2.

Zanon, S., 1926. Il terremoto di Venezia del 1° gennaio 1926. Rivista mensile della citta di Venezia, vol. V (9), pp. 383–392 (in Italian).

Zupan Hajna, N., Bosák, P., Pruner, P., Mihevc, A., Hercman, H., Horáček, I., 2020. Karst sediments in Slovenia: plio-quaternary multi-proxy records. Quat. Int. 546, 4–19. https://doi.org/10.1016/j.quaint.2019.11.010.

Zupančič, N., Šebela, S., Miler, M., 2011. Mineralogical and chemical characteristics of black coatings in Postojna Cave system. Acta Carsol. 40 (2), 307–317.

Internet sources

https://krog.sta.si/2588193/leto-2018-najboljse-v-zgodovini-turisticnega-obiska-postojnske-jame. (Accessed 11 December 2019).

http://www.esrl.noaa.gov/gmd/ccgg/trends. (Accessed 14 November 2020).

https://www.postojnska-jama.eu/en/media-centre/news-and-press-releases/postojna-cave-welcomes-its-38-millionth-visitor/. (Accessed 10 December 2020).

http://www.rtvslo.si/tureavanture/novice/park-postojnska-jama-podrl-mejnik-14-000-obiskovalcev-dnevno/463285. (Accessed 10 December 2020).

https://www.postojnska-jama.eu/sl/medijsko-sredisce/novice/rekordno-leto-v-postojnski-jami/. (Accessed 15 November 2020).

https://www.rtvslo.si/tureavanture/podobe-slovenije/postojnska-jama-leto-koncuje-z-zivimi-jaslicami/475436. (Accessed 10 September 2020).

https://siol.net/posel-danes/novice/zaradi-smrtonosnega-virusa-kitajci-odpovedujejo-potovanja-v-slovenijo-518560. (Accessed 6 November 2020).

https://www.rtvslo.si/tureavanture/novice/39-milijonti-obiskovalec-postojnske-jame-na-ogled-200-let-stara-vpisna-knjiga/489480. (Accessed 6 November 2020).

http://www.noviceznotranjske.net/park-postojnska-jama-s-14-000-obiski-v-enem-dnevu-znova-podrl-rekord/. (Accessed 14 November 2020).

https://whc.unesco.org/en/tentativelists/6072/. (Accessed 10 December 2020).

http://www.pisrs.si/Pis.web/pregledPredpisa?id=ZAKO2068. (Accessed 10 December 2020).

http://meteo.arso.gov.si/uploads/probase/www/climate/text/en/publications/OPS21_brosura_ENG.pdf. (Accessed 10 December 2020).

https://meteo.arso.gov.si/met/sl/archive/. (Accessed 10 December 2020).

http://www.i-s-c-a.org/objectives-and-purpose. (Accessed 6 November 2020).

https://www.jamarska-zveza.si/index.php/strokovne-sluzbe/ss-kataster-jam/ss-kj-prejeto-gradivo/78-kataster-arhiv/47-najdaljse-in-najgloblje-jame-v-sloveniji. (Accessed 6 November 2020).

Index

'*Note*: Page numbers followed by "f" indicate figures and "t" indicate tables.'

C

MIX
Papier | Fördert gute Waldnutzung
FSC® C083411

Kolibri 360 GmbH
Ferdinand-Jühlke-Straße 7
99095 Erfurt, Deutschland
produktsicherheit@kolibri360.de